Health Effects of Pesticides

Health Effects of Pesticides

A K Srivastava ▪ C Kesavachandran

CRC Press
Taylor & Francis Group
Boca Raton London New York

CRC Press is an imprint of the
Taylor & Francis Group, an **informa** business

The Energy and Resources Institute

CRC Press
Taylor & Francis Group
6000 Broken Sound Parkway NW, Suite 300
Boca Raton, FL 33487-2742

First issued in paperback 2023

ISBN 13: 978-1-03-265384-6 (pbk)
ISBN 13: 978-0-367-17518-4 (hbk)
ISBN 13: 978-0-429-05821-9 (ebk)

DOI: 10.1201/9780429058219

Print edition not for sale in South Asia (India, Sri Lanka, Nepal, Bangladesh, Pakistan or Bhutan)

Publisher's Note
The publisher has gone to great lengths to ensure the quality of this reprint but points out that some imperfections in the original copies may be apparent.

Library of Congress Cataloging in Publication Data
A catalog record has been requested

Visit the Taylor & Francis Web site at
http://www.taylorandfrancis.com

and the CRC Press Web site at
http://www.crcpress.com

PREFACE

Science and technology have been two important drivers for economic development in India. They have been responsible for increased production in both agricultural and industrial sectors and have fuelled the sociocultural and economic development of the country. The green revolution shepherded innovative technologies for control of pests besides other farming practices, resulting in significant increases in food production.

Pesticides were developed for killing life forms. The need to protect scarce and essential resources from other living things is truly prehistoric but it is only in the last century that toxic chemicals has come into widespread use. These chemicals kill the pests but their residues remain in the food and enter the human body along with it. They are sequestrated but human organs get affected by them. Pesticide poisoning may affect a single organ system or, alternatively, it may damage multiple organs. Detection of pesticide or its metabolites in body fluids requires a complex set-up, the lack of which results in misdiagnosis of delayed effects of pesticides. The current focus of medical therapeutics on microbial and/or nutritional causality of disease further complicates the issue.

Pesticides are useful in agricultural sector and in control of vector-borne diseases, but they may cause illness and death in humans. People most susceptible to direct contact with pesticides are subjects constantly exposed to pesticides either in the agricultural sector or in the pesticide manufacturing sector. Workers engaged in mixing, loading, and application of pesticides are exposed to pristine forms of pesticides and they may absorb these pesticides through their skin. This may result in both local effects on the skin and systemic poisoning through dermal absorption. Indirect contact with pesticides results from the ingestion of pesticide residues in food. This leads to an increased body burden of pesticides over a long duration and may or may not be associated with illness.

Depending on the specific biochemical mechanism of action, a pesticide may have widespread effects throughout the body, or it may cause a very limited change in physiological functioning of a particular organ. Innumerable factors play a role in disease production or detoxification and segregation/removal of the poison from the body.

Today, it is almost impossible for anyone to avoid frequent exposure to several different pesticides residues because of their extensive use. No wonder there is concern about possible adverse effects on human health arising from continual, long-term, low-level pesticide exposure. There is a particular concern regarding possible toxic effects in relation to carcinogenicity, mutagenicity, teratogenicity, neurobehavioural and neurotoxic damage, including allergic, and other immunoregulatory disorders.

This book is an endeavour to create awareness and share pertinent facts about pesticide usage across all stakeholders with an expectation that it will help in controlling the unbridled march of metabolic and the so-called 'lifestyle diseases'. The authors have addressed the issues relating to toxicity, known and probable adverse health impacts, exposure assessment studies, human studies reported from developing countries, acute and chronic toxicity profiles, current laws and legislations in India, and gap areas in this field.

The authors gratefully acknowledge the work and interest of our colleagues who initiated human studies on pesticide use among mango plantation workers at Epidemiology Section of IITR in mid-1980s and to those who continue to delve deeper into the area till date. The authors are thankful to the publisher, TERI Press, for the processing of the manuscript in time.

Dr A K Srivastava
Dr C Kesavachandran

CONTENTS

CHAPTER 1

INTRODUCTION

BRIEF HISTORY OF PESTICIDES

Any substance or a mixture of substances that prevents, destroys, or repels pests is termed as pesticide (EPA 2012b). It could be a chemical substance, a biological agent (such as a virus or bacterium), an antimicrobial, or a disinfectant. Pests include insects, plant pathogens, weeds, molluscs, birds, mammals, fish, nematodes (roundworms), and microbes that destroy property, spread diseases, or are vectors for diseases or cause nuisance.

The present-day Iraq, Turkey, Syria, and Jordan were the Fertile Crescent of Mesopotamia about 10,000 years ago. In this region, the practice of agriculture first began when a population of hunter/gatherers started collecting edible seeds (EPA 2012b). As settlement progressed, people in the region started cultivating wheat, barley, peas, lentils, chickpeas, bitter vetch, and flax. Eventually farming became the way of life (Unsworth 2010).

Archaeologists have found that along the banks of the Ganges rice was grown as a domesticated crop in the sixth millennium BC. Later, it extended to other areas. During the same time, in south-west Asia, barley, oats, wheat, lentil, and chickpea were cultivated. Evidence suggests that before the sixth millennium BC these cereals and legumes had been domesticated in north-west India. Other crops grown 3000–6000 years ago include oilseeds such as sesame, linseed, safflower, mustards, and castor; legumes such as mung bean, black gram, horse gram, pigeon pea, field pea, grass pea (*khesari*), and fenugreek; fibre crops such as cotton; and fruits such as jujube, grapes, dates, jackfruit, mango, mulberry, and black plum. Animals including livestock, sheep, goats, donkeys, dogs, pigs, and horses were also domesticated during the period (Mehra 1997).

In Africa, about 7500 years ago rice and sorghum were grown in the Sahel region. In China too, rice and millet were domesticated.

Potato was domesticated in South America. Around 3500 BC, corn and squash were cultivated in Mesoamerica. The Native Americans in 2500 BC cultivated sunflower apart from other crops.

Birds, mammals, microbes, insects, weeds, and so on have always been a threat to crops. These pests and diseases have affected the crop yield, with the threat of famine always looming. Even today, despite the advancements made in agricultural science, a considerable quantity of food products are destroyed due to pests (Peshin 2002).

To overcome the problems caused by pests and diseases, humans turned towards the use of pesticides. According to records, about 4500 years ago, Sumerians were the first to use sulphur compounds to control insects and mites. Ancient Romans burnt sulphur to kill insect pests and used salt to control weeds. Mercury and arsenical compounds were used by the Chinese about 3200 years ago to control body lice (Unsworth 2010).

In 1500 BC, Egyptians produced insecticides against lice, fleas, and wasps. *Geoponika* is a collection of 20 books on agricultural lore. It is a Greek agricultural encyclopedia that lists insecticides used during the tenth century and mentions the insecticidal properties of bay, asafoetida, elder, cumin, hellebore, aquill, cedar, absinthe, and pomegranate. Some of the pests mentioned in recorded history are locust swarms in the list of kosher animals in the Bible, cave paintings in Tassili n'Ajjer (Algeria) show crop infestations, Egyptian papyrus documents about pests; Holmer (800 BC) recognized usefulness of burning fields to control locusts[1]. Furthermore a concoction of honey and arsenic was used to control ants in the 1600s (Delaplane 1996). The London Horticulture Society, in 1821, suggested that in order to prevent mildew on peaches, sulphur should be used. Arsenic was applied to potato crops in the United States when Colorado beetle invaded the crops in 1867. In 1892, potassium dinitro-2-cresylate, the first synthetic pesticide, was marketed in Germany.[2] During World War II, inorganic and biological substances such as Paris green (copper acetoarsenite), lead arsenate, calcium arsenate, selenium compounds, lime–sulphur, pyrethrum, thiram, mercury, copper sulphate, derris, and nicotine were used, but their amounts and frequency of use were limited, and most pest control practices employed cultural methods such as rotations, tillage, and manipulation of sowing dates.[3] There was an increased application of pesticides after World War II. Several new

---

[1] Details available at <http://adamoliverbrown.com/wp-content/uploads/2012/01/2-History-of-Pesticides-6-slides.pdf>, last accessed on 13 March 2010

[2] Details available at <www.chm.bris.ac.uk/webprojects2000/aroshier/history.html>, last accessed on 13 March 2010

[3] Details available at <www.pollutionissues.com/Na-Ph/Pesticides.html>, last accessed on 13 March 2010

pesticides, such as DDT, aldrin, BHC, dieldrin, endrin, and 2,4-D were introduced.

Dichlorodiphenyl-trichloroethane (DDT), the first synthetic organochlorine insecticide, was discovered in Switzerland in 1939. DDT was especially preferred for its broad-spectrum activity against insect pests that impacted agriculture and human health. It was used extensively against head and body lice, human disease vectors, and agricultural pests till the 1970s. Benzene hexachloride (BHC) and chlordane were developed during World War II and toxaphene (and heptachlor) slightly later.[4] 2,4-D, an inexpensive and effective insecticide, was used in grass crops, such as corn, to control weeds (Delaplane 1996). These new chemicals became enormously popular as they were inexpensive and effective.

By the late nineteenth century, farmers in the United States were using Paris green, calcium arsenate, nicotine sulphate, and sulphur to control insect pests in field crops, but often results were poor because of the primitive chemistry and application methods (Delaplane 1996). Users not aware of the harmful effects of pesticides started using them generously with the aim of sterilizing the pests (Delaplane 1996). Some pests, under constant chemical pressure, became genetically resistant to pesticides (Figure 1). Also, pesticides started harming non-target plants and animals, and pesticide residues began appearing in unexpected places. It was not until the publication of *Silent Spring* by Rachel Carson (Carson 1962) in 1962 that people became aware of the harmful effects of pesticides. Carson, in her book, highlighted the risks and consequences of unhibited use of pesticides (Delaplane 1996).

Researchers, as a consequence, focused their research towards finding pest-specific pesticides and cropping methods that were not pesticide-centric. Scientists in the 1960s started working on an alternate approach, known as integrated pest management (IPM) (Delaplane 1996). The primary objective of IPM was to keep pests at economically insignificant levels

- using crop production methods that inhibit pests,
- encouraging the use of beneficial predators or parasites that attack pests, and
- timing pesticide applications to coincide with the most susceptible period of the pest's life cycle.

---

[4] Details available at <www.pollutionissues.com/Na-Ph/Pesticides.html>, last accessed on 13 March 2010

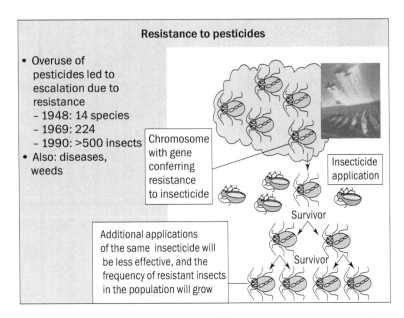

Figure 1 Increase in resistance to pesticides in subsequent generations

The objective of IPM is not necessarily to eradicate pesticides. Pests can be tolerated at certain low levels (Delaplane 1996). IPM also recognizes the fact that complete eradication of pests may affect the predators or parasites that survive on them (Delaplane 1996). The main aim of IPM is to either make pesticide usage more effective or reduce the overall use of pesticides (Delaplane 1996).

Presently, more and more pesticides are being derived from "natural" pesticides (Delaplane 1996). Synthetic pyrethroid is an insecticide modelled after naturally occurring plant-derived substances called pyrethrums that have been used as insecticides for centuries. Synthetic pyrethroids were introduced in the 1960s and include tetramethrin, resmethrin, fenvalerate, permethrin, lambda-cyhalothrin, and deltamethrin. These are quite commonly used in agriculture. Synthetic pyrethroid insecticides act similar to hormones that affect insect growth. They are broad-spectrum insecticides, do not bio-accumulate, and do not affect the non-target mammals. However, they are very toxic to aquatic invertebrates and fish. Synthetic pyrethroid insecticides and similar ones using bacteria, viruses, or other natural pest control agents are called "biorational" pesticides (Delaplane 1996).[5]

---

[5] Details available at <www.pollutionissues.com/Na-Ph/Pesticides.html>, last accessed on 13 March 2010

At present, 1600 types of pesticides are available in the market. Approximately, 4.4 million tonnes of pesticides are used every year, at a cost of more than $20 billion. The United States accounts for 25% of the pesticide market in the world.[6]

India is a largely agrarian society with more than 100 million families dependent on agriculture for their livelihood. Despite liberalization, Indian economy still relies on the success of agricultural production, and 35%–40% of India's national income comes from agricultural sources (Padole and Thakkar 2013). India started using pesticide in 1948 when DDT was imported for controlling malaria and BHC for controlling locust. The production of DDT started in India in 1952. By 1958, more than 500 tonnes of pesticides were produced in the country. At present, India produces approximately 85,000 tonnes of pesticides and around 145 pesticides are registered in the country (Kumar and Kumar 2007). In Asia, India is the most prominent manufacturer of basic pesticides and it ranks 12th in pesticide production worldwide. Of the total pesticides consumed in India, insecticides comprise 75%, fungicides 12%, and herbicides 10%. About 50% of the total pesticide used is sprayed on cotton field, 17% on rice, and 13% on vegetables and fruits (Indra, Bellamy, and Shyamsundar 2007).

Ancient Indian scriptures reveal that sages in the Vedic period possessed knowledge of plant diseases and pests. They developed eco-friendly methods of crop cultivation and organic agents as pesticides (Nath 2012). Other ancient scriptures which discuss plant protection practices include *Rigveda*, *Atharvaveda*, Kautilya's *Arthashastra*, Amarsimha's *Amarakosha*, Patanjali's *Mahabhasya*, *Krishi-Parashara*, *Sangam* literature of Tamils, *Agnipurana*, Varahamihira's *Brihat Samhita*, *Kashyapiyakrishisukti*, Surpala's *Vrikshayurveda*, Someshwara Deva's *Manasollasa*, and so on. These books provide a clear idea about plant protection practices used in India since millennia (Nath 2012).

In 600 AD, Varahamihira described about the treatment of trees. He mentioned in his book that when trees are exposed to cold weather, strong winds, and hot sun, they become vulnerable to diseases (Nath 2012). Their leaves become pale white, sprouts scanty and unhealthy, and branches dry with oozing of sap. Surpala (1000 AD) compiled a text called *Vrikshayurveda* (vriksha = trees, and ayurveda = science) and described the concept as applied to trees. It appears that the basic knowledge of *Vrikshayurveda* remained mostly confined to scholars. Some of Surpala's *Vrikshayurvedic*

---

[6] Details available at <www.pollutionissues.com/Na-Ph/Pesticides.html>, last accessed on 13 March 2010

verses dealing with plant protection prescriptions, translated by Sadhale and the translated stanza numbers (only 187–191) as given in Nath (2012), are as follows:

- **187:** The diseases of the *kafa* type can be overcome with bitter, strong, and astringent decoctions made out of *panchamula* (roots of five plant species—*sriphala, sarvatobhadra, patala, ganikarika,* and *syonaka*) with fragrant water.
- **188:** For warding off all *kafa* type of diseases, the paste of white mustard should be deposited at the root and trees should be watered with a mixture of sesame and ashes.
- **189:** In the case of trees affected by the *kafa* disease, earth around the roots of the trees should be removed and replaced with fresh, dry earth.
- **190:** A wise person should treat all types of trees affected by the *pitta* type of diseases with cool and sweet substances.
- **191:** When watered by the decoction of milk, honey, *yastimadhu,* and *madhuka,* trees suffering from *pitta* type of diseases get cured.

Ever since the Vedic period, the benefits of neem have been known in India. Apart from being used to control aphids, whitefly, moth larvae, scale, Japanese beetles, spider mites, it was also used for treating mites (Pragya 2011). Hence, it is effective both as a pesticide and as a miticide. Kautilya in his *Arthashastra* has mentioned about the domestic uses of neem (Padole and Thakkar 2013).

Crops, biotic and abiotic, can be protected by following the procedures mentioned in the ancient documents of Indian subcontinent. The methodologies mentioned in Vrikshayurvedic for biotic and abiotic stresses, when adopted, will prove economically beneficial to farmers. The community will also benefit by practising pesticide-free farming as it will ensure agricultural sustainability (Nath 2012).

TYPES OF PESTICIDES AND THEIR USAGE

Classification of pesticides can be based on many factors. They can be grouped as inorganic, synthetic, or biopesticides (AMA 1997). Based on the target organism, chemical structure, and physical status too, pesticides can be classified (AMA 1997). Biopesticides can be microbial and biochemical pesticides (EPA 2012a). Pyrethroids, rotenoids, nicotinoids, and a fourth group that includes strychnine and scilliroside can be placed under plant-derived pesticides or botanicals (Kamrin 1997). Herbicides, insecticides, fungicides, rodenticides, pediculicides, and biocides are the subclasses of pesticides.

Pesticides can also be classified in terms of chemical families, such as organochlorines, organophosphates, or carbamates. After the end of the World War II, aldrin and dieldrin, two cyclodiene organochlorines, were introduced, followed by endrin, endosulfan, and isobenzan. These cyclodiene organochlorine insecticides affected an insect's nervous system, leading to malfunction, tremors, and death. Organochlorines have harmful effects on the ecology. As they are relatively insoluble, their traces can be found in soils and aquatic sediments. They bioconcentrate in the tissues of invertebrates and vertebrates from their food, move up trophic chains, and affect top predators. As organochlorines possessed the properties of persistence and bioaccumulation, their use was banned in the industrialized nations from 1973 to late 1990s. Yet they continued to be used in the developing countries.[7] DDT is an example of organochlorine hydrocarbons. The organochlorine hydrocarbons can be further classified into dichlorodiphenylethanes, cyclodiene compounds, and other related compounds (Kamrin 1997).

Organochlorines were largely replaced by organophosphates and carbamates. These two types of pesticides act by inhibiting the enzyme acetylcholinesterase, allowing acetylcholine to transfer nerve impulses indefinitely and causing a variety of symptoms such as weakness or paralysis. Organophosphate insecticides were developed in Germany during World War II. Originating from compounds developed as nerve gases, these chemicals were produced as insecticides, such as tetraethyl pyrophosphate, parathion, demeton, methyl schradan, phorate, diazinon, disulphoton, dimethoate, trichlorophon, and mevinphos. Organophosphate insecticides have the same effect in mammals as in insects. They inhibit the cholinesterase enzyme that breaks down the neurotransmitter acetylcholine at the nerve synapse, thereby blocking impulses and causing hyperactivity and tetanic paralysis of the insect, followed by death. Some of the organophosphate insecticides are systemic in plants and animals. They do not have significant environmental impacts, and they are neither persistent nor bioaccumulate in animals.[8]

In comparison to organophosphates, carbamates are less toxic (Kamrin 1997). In 1951, Geigy Chemical Company introduced carbamates.[9] These

---

[7] Details available at <www.pollutionissues.com/Na-Ph/Pesticides.html>, last accessed on 13 March 2010

[8] Details available at <www.pollutionissues.com/Na-Ph/Pesticides.html>, last accessed on 13 March 2010

[9] Details available at <www.chm.bris.ac.uk/webprojects2000/aroshier/history.html>, last accessed on 13 March 2010

insecticides are broad spectrum with moderate toxicity and persistence. They neither generally bioaccumulate nor have major environmental impacts.[10] Carbaryl was the first carbamate insecticide. Aldicarb, methiocarb, methomyl, carbofuran, bendiocarb, and oxamyl are the other examples of carbamates. Carbaryl affects the nervous transmissions in insects. It inhibits the cholinesterase activity by blocking acetylcholine receptors. Thiocarbamate and dithiocarbamates are subclasses of carbamates. Prominent families of herbicides include pheoxy and benzoic acid herbicides (e.g. 2, 4-D), triazines (e.g. atrazine), ureas (e.g. diuron), and chloroacetanilides (e.g. alachlor). Phenoxy compounds tend to selectively kill broadleaved weeds rather than grasses. The phenoxy and benzoic acid herbicides act similar to plant growth hormones and promote growth of cells without normal cell division, crushing the plant's nutrient transport system (Kamrin 1997). Triazines interfere with photosynthesis (Kamrin 1997). Many commonly used pesticides are not included in these families, including glyphosate.

In the last few decades, the emphasis has been on developing pesticides that neither bioaccumulate nor are persistent. These new classes of insecticides are of the following forms: juvenile hormone mimics, synthetic versions of insect juvenile hormones that act by preventing immature stages of the insects from molting into an adult, avermectins, and natural products produced by soil microorganisms. *Bacillus thuringiensis* toxins are proteins produced by a bacterium that is pathogenic to insects. When they are activated in the insect gut, they destroy the selective permeability of the gut wall. The first strains were toxic only to *Lepidoptera,* but subsequently strains toxic to flies and beetles have been developed. *B. thuringiensis* was incorporated into plants genetically.[11] Some important classes of pesticides are listed in Table 1.

POTENTIAL HEALTH HAZARDS

Although pesticides have several benefits, there are also several disadvantages associated with their use, such as potential toxicity to humans and other animals. According to the Stockholm Convention on Persistent Organic Pollutants, 10 of the 12 most dangerous and persistent organic chemicals include pesticides (Gilden, Huffling, and Sattler 2010).

There is a growing concern worldwide regarding the presence of pesticide residues in drinking water and food, which are often suspected

---

[10] Details available at <www.pollutionissues.com/Na-Ph/Pesticides.html>, last accessed on 13 March 2010

[11] Details available at <www.pollutionissues.com/Na-Ph/Pesticides.html>, last accessed on 13 March 2010

Table 1 Major classes of pesticides

Type of pesticide	Target pest group
Acaricide	Mites, ticks, spiders
Antimicrobial	Bacteria, viruses, other microbes
Attractant	Attracts pests for monitoring or killing
Avicide	Birds
Fungicide	Fungi
Herbicide	Weeds
Insecticide	Insects
Molluscicide	Snails and slugs
Nematicide	Nematodes
Piscicide	Fish
Predacide	Vertebrate predators
Repellant	Repel pests
Rodenticide	Rodents
Synergist	Improves performance of another pesticide

Source Delaplane (1996)

of hampering endocrine activities and being carcinogenic. Despite international and national regulatory agencies implementing stringent regulations, pesticide residues have been increasingly reported in human foods, both imported and home produced.[12]

In the past five decades, many human diseases and deaths have occurred because of pesticide exposure. Around 20,000 deaths have been reported to have occurred annually as a result of pesticide use. Some of these deaths have been in the form of suicides, while many involve some form of accidental exposure to pesticides, particularly among farmers and spray operators in developing countries. Farmers and spray workers fail to take adequate precautions while handling pesticides or they do not use protective clothing and equipment. Thousands of people have died or become ill as a result of accidental exposure to pesticides. The mishap that occurred in Bhopal claimed more than 5000 deaths. It resulted from exposure to accidental emissions of methyl isocyanate from a pesticide factory.[13] The health issues are discussed in detail in Chapter 3.

---

[12] Details available at <www.pollutionissues.com/Na-Ph/Pesticides.html>, last accessed on 13 March 2010

[13] Details available at <www.pollutionissues.com/Na-Ph/Pesticides.html>

REFERENCES

AMA (American Medical Association). 1997. Educational and informational strategies to reduce pesticide risks. *Preventive Medicine* 26(2): 191–200

Carson, R. 1962. *Silent Spring*. Boston: Houghton Mifflin Company

Delaplane, K. S. 1996. Pesticide usage in the United States: history, benefits, risks, and trends. Details available at <www.ipm.ncsu.edu/safety/factsheets/pestuse.pdf>, last accessed on 14 March 2010

EPA (Environmental Protection Agency). 2012a. Types of pesticides. Details available at <www.epa.gov/pesticides/about/types.htm>, last accessed on 23 March 2010

EPA (Environmental Protection Agency). 2012b. What is a pesticide? Details available at <www.epa.gov/pesticides/about/index.htm>, last accessed on 13 January 2010

Gilden, R. C., K. Huffling, and B. Sattler. 2010. Pesticides and health risks. *Journal of Obstetric, Gynecologic, and Neonatal Nursing* 39(1): 103–110

Indra, D. P., R. Bellamy, and P. Shyamsundar. 2007. Facing hazards at work: agricultural workers and pesticide exposure in Kuttanad, Kerala. *South Asian Network for Development and Environmental Economics* 19: 1–4

Kamrin, M. A. (ed). 1997. *Pesticide Profiles: toxicity, environmental impact, and fate*. New York: CRC Press

Kumar, M. and A. Kumar. 2007. Application and health effects of pesticide commonly used in India. Details available at <www.eco-web.com/edi/070526.html>, last accessed on 1 July 2010

Mehra, K. L. 1997. Biodiversity and subsistence changes in India: the Neolithic and Chalcolithic Age. *Asian Agri-History* 1: 105–126

Nath, P. 2012. Management of biotic stresses of crop plants through vrikshayurveda techniques. Details available at <www.fnu.ac.fj/images/stories/CAFF/dean_seminor.pdf>, last accessed on 23 March 2010

Padole L. and P. Thakkar. 2013. Neem use and potential in agriculture. Details available at www.neemfoundation.org/neem-articles/neem-in-organic-farming/agricultural-usepotential.html, accessed on 11 April 2010

Peshin, R. 2002. Economic benefits of pest management. In *Encyclopedia of Pest Management*, pp. 224–227. New York: Marcel Dekker

Pragya, T. 2011. Neem oil insecticide. Details available at <www.buzzle.com/articles/neem-oil-insecticide.html>, last accessed on 13 March 2010

Unsworth, J. 2010. History of pesticide use. Details available at <http://agrochemicals.iupac.org/index.php?option=com_sobi2&sobi2Task=sobi2Details&catid=3&sobi2Id=31>, last accessed on 13 March 2010

Ware, G. W. 1994. *The Pesticide Book*, 4th edn. California: Thomson Publications

CHAPTER 2

ACUTE AND CHRONIC TOXICITY

TOXICITY

The degree to which an organism can be harmed by a substance is known as toxicity. Toxicity can affect the health of the organism as it has the ability to alter the normal physiological, biochemical, and pathological conditions. The scope of these effects varies from headache, coma, convulsions, to even death (Nesheim, Fishel, and Mossler 2012).

Animals, such as dogs, rabbits, and mice are typically employed for carrying out the process of pesticide testing. The primary aim of testing is to find out the toxicity type and dosage required for determining a toxic reaction (Nesheim, Fishel, and Mossler 2012). Some of the effects caused are not necessarily harmful in the long run and are reversible. This can also be ensured with a prompt medical assistance. However, there are toxicants whose effects are irreversible. Various international bodies are involved in the development of guidelines for testing of pesticides.

Organization for Economic Cooperation and Development (OECD) has developed guidelines for chemical testing. The OECD's guidelines contain internationally accepted methods for pesticide testing. These methods are employed by the industry, government, and independent laboratories and are used to figure out safety of chemicals and their preparations. These guidelines also cover industrial chemicals and pesticides (OECD 2012). Toxicity can be categorized as acute and chronic on the basis of the number of exposures to poison and time taken for developing toxic symptoms. Exposure is of the short duration in the case of acute toxicity and results can be observed within a short period of time. Chronic toxicity is the result of repeated or long-term exposure to a poison. In the chronic toxicity, adverse results

Table 1 Type of toxicity

Type of toxicity	Time for symptoms to develop
Acute	Immediate (minutes to hours)
Chronic	One week to years

Source   Nesheim, Fishel, and Mossler (2012)

are observed after a considerably long time (Nesheim, Fishel, and Mossler 2012) (Table 1).

Test animals are subjected to numerous dosages of the active ingredient and its formulated products for determining the pesticide's toxicity. The pest is controlled by the chemical component of the pesticide called the active ingredient (Hock and Lorenz 2006). For making pesticide users aware of the acute toxicity of a pesticide, different marking labels are used. There are four types of marking labels: highly toxic, moderately toxic, slightly toxic, and relatively non-toxic (Fischel, Nesheim, and Mark 2011).

Acute Toxicity

Acute toxicity can be defined as a chemical's or pesticide's capacity to bring systemic damage as a consequence of one-time exposure to a fairly large amount of the chemical (Nesheim, Fishel, and Mossler 2012).

Headache, rash, vision problems, and breathlessness are common symptoms found in farmers exposed to acute pesticide poisoning. Acute toxicity, its measures, and warnings are listed in Table 2.

Acute toxicity tests (short term) are typically employed for determining the effects of exposure to high amounts of chemicals. The end result indicates the level of lethality (USEPA 1994). On the contrary, the chronic toxicity tests are typical long-term tests conducted to find the effects of exposure to comparatively less toxic concentrations. The end result pinpoints to the sub-fatal effect (for example, reproduction, growth) or both lethality and sub-lethal effects (USEPA 1994).

The acute toxicity of pesticide can cause injury to a person or an animal with single exposure of short duration. This exposure can occur through different routes, such as dermal (skin), inhalation (lungs), oral (mouth), and eyes. Test animals are examined on the basis of their dermal toxicity, inhalation toxicity, and oral toxicity for finding out the acute toxicity. Eye and skin irritations are also considered while examining the test animals (Nesheim, Fishel, and Mossler 2012; Hock and Lorenz 2006).

Table 2 Acute toxicity measures and warnings

Categories		Signal word	Categories of acute toxicity			Oral lethal dose
			LD_{50}	LD_{50}	LC_{50}	
			Oral (mg/kg)	Dermal (mg/kg)	Inhale (mg/L)	
I	Highly toxic	Danger poison (skull and crossbones)	0–50	0–200	0–0.2	A few drops to a teaspoonful
II	Moderately toxic	Warning	50–500	200–2,000	0.2–2.0	Over a teaspoonful to one ounce
III	Slightly toxic	Caution	500–5,000	2,000–20,000	2.0–20	Over one ounce to one pint
IV	Relatively non-toxic	Caution (or no signal word)	5,000+	20,000+	20+	Over one pint to one pound

Note: Probable for a 68 kg person
Source Nesheim, Fishel, and Mossler (2012)

LD_{50} is the term usually used while referring to acute toxicity. LD stands for lethal dose and 50 implies that the 50 per cent of the animals on whom the test was done were severely affected by the given dose under the controlled conditions (Fischel, Nesheim, and Mark 2011). LD_{50} is expressed in terms of milligrams (mg) of pesticide per kilogram (kg) of the body weight and has been accepted as a standard for measuring acute toxicity (USEPA 2012). Test animals are observed for a particular period of time after giving chemical's dose either orally or through injection. The dose required to kill 50 per cent of the test animals is expressed by LD_{50} (USEPA 2012). LD_{50} has made it possible to compare the relative toxicities of pesticides because its values are standard measurements (USEPA 2012). The value of LD_{50} confirms the toxicity of the pesticide. The lower the value of the LD_{50}, the more toxic the pesticide is and vice versa (USEPA 2012). Active ingredient forms the basis for expressing LD_{50} values. For example, if a pesticide contains 50 per cent active ingredient, two parts of its material will be required for making one part of the active ingredient. However, in some cases, toxicity may differ if other chemicals are mixed along with the active ingredient in the production of the pesticide (Fischel, Nesheim, and Mark 2011).

The acute toxicity test can estimate the concentration of the medium required to kill 50 per cent of test animals. This is called the lethal concentration (LC_{50}) and is the median lethal concentration. When an acute toxicity test is reported as the LC_{50} value, the duration of the test, the test species, and the life cycle stage of the test species are specified by the test result. LC_{50} can be treated as point estimates, which are estimates of the effects of particular concentrations of contaminants. We can determine coefficients of variation for LC_{50} from these estimates (USEPA 1994). The LC_{50} values are expressed in milligrams per litre. The lower the value of LC_{50}, the more toxic is the pesticide and vice versa (Fischel, Nesheim, and Mark 2011).

> LD_{50} signifies the amount of the dose required for killing half of the test animals. On the other hand, LC_{50} is that concentration (in terms of water or gas) which can kill half of the test animals.

Data can be analysed to find out the dilution at which 50 per cent of the organisms showed the effect. This dilution is termed as median effective concentration (EC_{50}). When EC_{50} is reported by an acute toxicity test, the species used, duration of the test, and life cycle stage of the test species are also confirmed by the test results. EC_{50} is also a point estimate like its counterpart LC_{50} and therefore a coefficient of variations can be determined (USEPA 1994).

From statistical perspective, the lowest observed effect concentration (LOEC) is the highest dilution causing toxic effects. The lowest dilution at which no statistically significant effects are observed is known as no observed effect concentration (NOEC). Coefficients of variation for LOECs and NOAELs cannot be determined as they are not point estimates (USEPA 1994).

The no observed adverse effect level (NOAEL) can be evaluated too in course of toxicity testing studies. This is the highest dose at which no adverse effects result (Figure 1) (Tharp 2008). Dose–response curve is shown in Figure 1 depicting NOAEL–LD_{50} relationship. Suicidal, accidental, and other pesticide poisonings are discussed in other chapters of the book.

Chronic Toxicity

To find out chronic toxicity of a pesticide, test animals are subjected to long-term exposure to the active ingredient of pesticide. Chronic effect is the term which signifies harmful effects resulting from exposure to

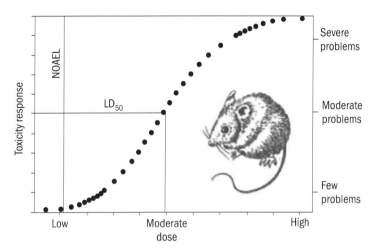

Figure 1 Dose–response curve indicating the NOAEL in relation to LD_{50}
Source Tharp (2008)

small doses of pesticides repeated over time. Birth defects, development of tumours, and nerve disorders are some of the chronic effects that result from pesticide exposure. To find out the acute toxicity of a pesticide in a laboratory is comparatively simpler than finding chronic toxicity (Hock and Lorenz 2006).

Sub-lethal effects can produce structural abnormalities such as endocrine disruption, growth reduction, nerve function impairment, lack of motility, behavioural changes, and the development of terata (Fischel, Nesheim, and Mark 2011; USEPA 1994).

To determine chronic toxicity estimations, the time period for which animals are exposed to sub-lethal doses can vary from months to several years. The following are some of the reactions for which the animals are assessed (Tharp 2008):

(i) Carcinogenesis – cancer

(ii) Teratogenesis – birth defects

(iii) Mutagenicity – genetic changes

(iv) Oncogenicity – tumours

(v) Reproductive – reproductive problems

(vi) Nervous – nerve damage

(vii)Immune – allergic reactions

It is hard to anticipate the precise doses and exposure duration required for the development of toxic effects (Tharp 2008). LOEC, EC_{50}, and

NOEC are some of the other approaches for measuring chronic toxicity (USEPA 1994). Differentiation of the products is done as per their LD_{50} and LC_{50} values. In other words, they are differentiated according to their relative acute toxicity. Pesticides can be classified as per their oral, dermal, and relative toxicities. Packaging label of the pesticides which fall under category I of toxicity must carry warning words "danger" and "poison" (in capital) in red colour along with clearly visible symbol of skull and crossbones (Hock and Lorenz 2006). For pesticide products of category I of toxicity, the value for acute oral LD_{50} can vary from a very small amount to 50 mg/kg. For example, even oral intake of a few drops of a material can be extremely harmful to a person weighing 68 kg. Pesticides which fall under category II of toxicity essentially should have "warning" labelled on their packaging. Pesticides which belong to categories III and IV (less toxic or non-toxic) must have the "caution" word on their packaging labels. Values of acute oral LD_{50} in this class are higher than 500 mg/kg. Exposure of a trace amount or higher can prove deadly even for a person of 68 kg (Hock and Lorenz 2006). The toxicity categories (I to IV) are the outcome of the acute-oral, dermal, and inhalation tests. The acute LD_{50} or LC_{50} belongs to category I of the four toxicity categories. On the basis of the toxicity category in which the pesticide falls, a warning word is assigned to each pesticide and it is shown in Table 3 (Tharp 2008).

Table 4 summarizes the LD_{50} and LC_{50} values for the different routes of exposure and the effects on skin and eyes for the four toxicity categories and their associated signal word.

RISK ASSESSMENT OF LONG-TERM HAZARDS POSED BY PESTICIDE RESIDUES

Joint Meeting on Pesticide Residue (JMPR) is an expert body which conducts its meeting annually in order to mark the acceptable daily intakes (ADI) of a pesticide. JMPR reviews on the basis of the data collected on

Table 3   Signal words found on a product label

Signal word	Toxicity rating	Oral LD_{50} (mg/kg)	Lethal dose for 72 kg human
Danger	Highly toxic	0–50	Few drops to 1 teaspoon
Warning	Moderately toxic	50–500	1 teaspoon to 1 ounce
Caution	Slightly toxic	500–5,000	1 ounce to 1 pint
Caution	Relatively non-toxic	>5,000	Over 1 pint

Source   Tharp (2008)

Table 4 Toxicity categories for active ingredients

Routes of exposure	Toxicity categories			
	I	*II*	*III*	*IV*
Oral LD_{50}	Up to and including 50 mg/kg	50–500 mg/kg	500–5,000 mg/kg	>5,000 mg/kg
Inhalation LC_{50}	Up to and including 0.2 mg/L	0.2–2 mg/L	2–20 mg/L	>20 mg/L
Dermal LD_{50}	Up to and including 200 mg/kg	200–2,000 mg/kg	2,000–20,000 mg/kg	>20,000 mg/kg
Eye effects	Corrosive corneal opacity not reversible within 7 days	Corrosive corneal opacity not reversible within 7 days; irritation persists for 7 days	No corneal opacity; irritation reversible within 7 days	No irritation
Skin effects	Corrosive	Severe irritation at 72 h	Moderate irritation at 72 h	Mild or slight irritation at 72 h
Signal word	Danger poison	Warning	Caution	Caution

Source Hock and Lorenz (2006)

biochemical, metabolic, pharmacological, and toxicological properties of the pesticide, obtained from carrying out experiments on animals and observations in humans. The type of effect, the severity or reversibility of the effect, and the problems of inter- and intra-species variability are encapsulated in a security factor which is applied to the NOAEL to find out the ADI for humans. This evaluation acts as hazard characterization step in the risk assessment of long-term hazards posed by the residuals of the pesticide (WHO 1997).

It has become a matter of occasional concern that exposure to residues of more than one pesticide in food can result in adverse health effects. The importance of pesticide interactions was recognized in JMPR of 1981 (WHO 1997). It was confirmed that no major evidence was found which can compel to change the general approach in ADIs' estimation. However, the topic was re-raised in 1996 (WHO 1997) and this time it was well established that interactions between pesticide residues, other dietary constituents, and environmental contaminants could occur. It was also inferred that these interactions depend on numerous factors. Factors could be chemical and physical nature of the substances, the amount of dose, and conditions of exposure (WHO 1997).

The WHO Core Assessment Group and FAO Panel of Experts on Pesticide Residues in Food and Environment form the present JMPR. It has been adjudged as a successful model after its collaboration with WHO. Experts from government and academic institutions comprise a major part of JMPR. These experts act as independent, internationally recognized specialists carrying out the work in a personal capacity. They do not act as the mere representatives of their governments (WHO 2011). The WHO core assessment group does the job of reviewing pesticide toxicological and related data. It is also involved in figuring out estimates for NOAELs of pesticides and ADIs of their residues in food for humans. Also, based on the data and situation, acute reference doses (ARfDS) are calculated and other toxicological criteria such as non-dietary exposures are characterized by the group (WHO 2011).

Residue Levels

The toxicity of the active ingredients and its metabolites, evaluated by WHO core assessment group, is considered for assessing the risk posed by residues on human health. The maximum residue levels are recommended to the Codex Committee on Pesticide Residue (CCPR) for consideration to be adopted by the Codex Alimentarius Commission (CAC) (WHO 2011).

For estimating intake of pesticide residues, many indices of residue levels are utilized. Maximum residual level (MRL) is one example. It is expressed in mg/kg and represents the maximal concentration of a pesticide recommended by CAC as the permissible level in food products and animal feeds. Pesticide residue levels that would be expected from the FAO panel responsible for reviewing pesticide use patterns estimate the levels from a collection of nationally generated data. Both MRL and ADI are not assumed as permanent standards. A group comprising internationally renowned experts sets up MRL and ADI as per the data available at the time of evaluation (WHO 1997; WHO 2011).

Exposure Assessment

In order to find out the dietary intake of a pesticide residue in a given food, the residue level in the food is multiplied by the amount of the food consumed. Then by adding the intakes of all foods containing the residue, the total intake of the pesticide residue is calculated. Exposure to a pesticide residue present or probably present in drinking water or food whose MRL is unavailable is also taken into account if such information is available. The estimated dietary intake of pesticide residues resulting from application of a pesticide and other sources should be less than its established ADI. It is important to mention here that short-term intake of a pesticide above ADI is not a health hazard (WHO 1997).

Risk Characterization

The theoretical maximum daily intake (TMDI) is used as a standard at the international level for determining dietary intake. The international estimated daily intake (IEDI) is a better option for calculating dietary exposure in case the information is available. The risk characterization for TMDI and IEDI is based on the average adult weight of 60 kg. Precise average body weight is utilized for determining accurately the acceptable intake per person in some regions and even some countries. At the national level, national theoretical maximum daily intake (NTMDI) and the national estimated daily intake (NEDI) are employed for similar purposes (WHO 1997). The WHO (1997) recommended approach to dietary intake assessments at the international and national levels is presented in Figure 2.

Theoretical maximum daily intake

Codex MRLs are suitable for making first estimate of the pesticide intake which is termed as theoretical maximum daily intake (TMDI). Since the actual levels found in most foods are below the corresponding MRLs, the estimate should be employed only to separate those pesticides which lead

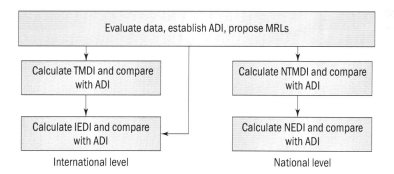

Figure 2 Scheme for the assessment of dietary intake of pesticide residues for long-term hazards
Source WHO (1997)

to no harmful effects following long-term intake from those pesticides that need further consideration (WHO 1997).

The calculation of TMDI is done by multiplying the established or proposed Codex MRLs by the estimated average daily regional consumption for each food commodity and then adding the products. This is given by the following relation:

TMDI = ΣMRL$_i$ × F_i

where

$\quad$ MRL$_i$ = Maximum residue limit for a given food commodity

$\quad\quad F_i$ = Per capita GEMS/food regional consumption of that food commodity

The TMDI is an overestimate of the true pesticide residue intake owing to the following reasons:

(i) Only a proportion of a specific crop is treated with a pesticide

(ii) Most treated crops contain residues at harvest below the MRL

(iii) Amounts of residues get reduced during storage, preparation, commercial processing, and cooking

(iv) It is unlikely that every food for which an MRL is proposed will have been treated with the pesticide over the lifetime of the consumer

Hence, the belief that when the TMDI exceeds the ADI, the proposed MRLs for a pesticide are unacceptable is wrong. However, if ADI is not exceeded by TMDI, then there is least possibility for ADI to exceed. This holds true even for consumers with high intake, but with the condition that Codex MRLs cover the main uses of the pesticide (WHO 1997).

International estimated daily intake

The international estimated daily intake (IEDI) is responsible for incorporating correction factors to be applied at the international level. IEDI provides the best estimate of dietary intake by refining the intake estimate on the basis of available information (WHO 1997).

National theoretical maximum daily intake

The national theoretical maximum daily intake (NTMDI) can be utilized for confirming the TMDI and it incorporates factors available at the national level. NTMDI works only at the national level. NTMDI is a gross overestimation of exposure although it can be employed as a screening tool for estimating the dietary intake of pesticide residues. NTMDI is a conservative but scientifically sound exposure assessment method. National authorities can accept the MRLs even if the NTMDI incorporating Codex MRLs are below ADI. It is important to stress that countries should make full use of the available relevant data to refine dietary intake estimates, particularly when NTMDI values are more than the ADI (WHO 1997). While countries may use different approaches, the following formula is often applied:

$$\text{NTMDI} = \Sigma \text{MRL}_i \times F_i$$

where

MRL_i = Maximum residue limit (or national maximum limit) for a given food commodity

F_i = National consumption of that food commodity per person

PLAN OF ACTION FOR ACUTE PESTICIDE POISONINGS

The plan given below is based on the earlier report "Pesticides and personal safety", Purdue Pesticide Programs, Purdue University Cooperative Extension Services (Whitford, Edwards, Neal, *et al.* 1992).

A well-organized plan should be laid down for a pesticide user when it comes to dealing with a pesticide-related accident. Advance planning and preparation should be a routine practice. It should be made mandatory for all employees to know emergency procedures to be followed.

Contact Medical Personnel

The foremost thing to do in any poisoning emergency is to avoid further exposure and ensure that the victim is breathing. Following this, medical assistance should be sought.

Maintain Vital Signs

The victim should be provided with the first aid until the medical help arrives. Maintenance of vital signs is crucial and this may require application of cardiopulmonary resuscitation techniques. Respiratory failure is the primary cause of death of most victims of pesticide poisoning. There are high chances of recovery if the oxygen supply to the body can be maintained. Only a specialist (doctor) will be able to treat the victim properly with medication and necessary equipment. The medical personnel attending the victim must be provided with a copy of the pesticide label.

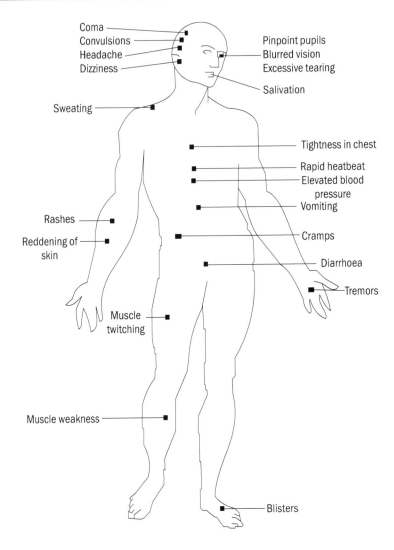

Figure 3 Symptoms of pesticide poisoning

Source Whitford, Edwards, Neal, *et al.* (1992)

Eliminate Further Contamination

Ingested pesticides

If a person swallows pesticide, he should be treated immediately rather than waiting for the symptoms to appear. It should be first confirmed whether vomiting can be induced or not. The pesticide label can help in deciding whether to induce vomiting. Vomiting should not be induced in case the victim is unconscious or convulsive. It has been observed that

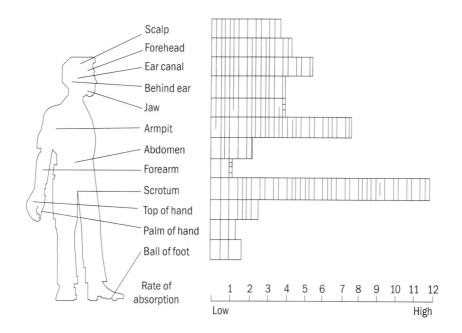

Figure 4 Specific body regions and their relative susceptibility to pesticide absorption

Source Whitford, Edwards, Neal, *et al.* (1992)

induced vomiting at the right time can mark the difference between life and death of a pesticide poisoned victim.

For inducing vomiting, ipecac syrup is very effective. Make sure that the victim is in a forward kneeling position or remains on his right side if he is lying down. This is done in order to avoid the vomits aspirating to his lungs. For removal of stomach contents, gastric lavage is another method performed by a physician. The procedure of the gastric lavage should be performed soon after the pesticide consumption has been reported and not after 2 h of the consumption. This is because within 2 h of its consumption, the pesticide will reach the intestine, and this calls for another method for poison removal. In that case physicians can opt for absorptive charcoals to prevent pesticide absorption from intestine and to facilitate the release of the pesticide through the faeces. The pesticide label should be checked prior to giving first aid as in some cases induced vomiting can worsen the situation. If pesticide's ingredient contains organic solvents or corrosives such as strong acids and bases, then induced vomiting should be avoided because these chemicals can cause permanent damage to sensitive tissues of oesophagus or the lungs during aspiration.

Pesticides on the skin

(i)   The pesticide should be washed off from the victim as early as possible so that further exposure and injury can be prevented.

(ii) Victim's clothes should be put off and the skin should be drenched with water. Hose, shower, faucet, and pond can be used for this purpose. Then clean skin and hair with soap and water. But care should be taken to ensure that skin is not injured during washing.

(iii) The victim should be wrapped in a blanket once the skin is dried.

Chemical burns of the skin

Taking immediate action is extremely important. Victim's clothes contaminated with chemicals should be immediately removed. Cold running water to be used for the skin wash. The affected area should be covered with a soft, clean cloth. Ointments, powders, greases, or any other drug should not be used in the case of chemical burns.

Pesticides in the eye

Wash the eye immediately, but it should be done gently. Eyelids should be kept open while washing with a gentle stream of water. Preferably water temperature should be same as body temperature. Continue washing for 15 min or extend it further. Chemicals and drugs should not be used as they tend to increase risk of injury.

Inhaled pesticides

A respirator should be worn while shifting the victim from an enclosed and contaminated area. Victim should be carried to the fresh air promptly. Tight clothing should be loosened. Artificial respiration can be applied if the victim is feeling breathlessness. Keep the victim as quiet as possible. Victim's breathing has to be watched in case he is convulsing. He should be protected from falling so that he does not end up injuring his head. Chin should be pulled forward to avoid blockage of the air passage by the tongue.

The victim can be covered with a blanket in order to prevent chilling. At the same time, it should be ensured that covering does not cause overheating of the body. As every pesticide is different, the product label should be verified to determine the requirements of personal protective equipment (PPE) for each pesticide; some general rules apply for selecting PPE based on the different toxicity categories are given in Table 5 (Hock and Lorenz 2006).

Table 5 Minimum PPEs and work clothing for pesticide handling activities

Routes of exposure	Toxicity categories			
	I	II	III	IV
Dermal toxicity or skin irritation potential	Coveralls worn over long sleeved shirt and long pants	Coveralls worn over short sleeved shirt and short pants	Long sleeved shirt and long pants	Long sleeved shirt and long pants
	Socks	Socks	Socks	Socks
	Chemical resistant footwear	Chemical resistant footwear	Shoes	Shoes
	Chemical resistant gloves	Chemical resistant gloves	Not available	Not available
Inhalation toxicity	Respiratory protection device	Respiratory protection device	Not available	Not available
Eye irritation potential	Protective eyewear	Protective eyewear	Not available	Not available

Source Hock and Lorenz (2006)

There are certain standard specifications required for PPEs which are used for protection of dermal, ocular, and respiratory parts as proposed by Tharp (2008) and given in Table 6. Always read the product label for specific instructions regarding personal protective equipment.

Table 6 Steps required for protection against the inherent acute and chronic toxicity

Action	Material	Phases	Protection of organ
Gloves	Nitrile, viton, neoprene, vinyl, rubber	All phases	Dermal
Apron	Chemically resistant	Mixing, handling	Dermal
Coveralls	Tyvek, Tychem, Rubber	All phases	Dermal
Goggles/ face shield	Chemically resistant	All phases	Ocular (eye)
Respirator	Variable	All phases	Respiratory/ ingestion

Source Tharp (2008)

REFERENCES

Fishel F. M., O. N. Nesheim, and A. M. Mark. 2011. *Toxicity of Pesticides.* Gainesville, FL: College of Agricultural Science, University of Florida IFAS Extension. Details available at http://edis.ifas.ufl.edu/pdffiles/PI/PI00800.pdf, last accessed on 4 April 2011

Hock, W. K. and E. S. Lorenz. 2006. *Toxicity of Pesticides.* University Park, PA: The College of Agricultural Sciences, The Pennsylvania State University. Details available at http://pubs.cas.psu.edu/FreePubs/pdfs/uo222.pdf, last accessed on 3 March 2013

Nesheim O. N., F. M. Fishel, and M. Mossler. 2012. Toxicity of pesticides. UF/IFAS EDIS Document PI-13. Details available at http://edis.ifas.ufl.edu/pi008, last accessed on 25 April 2014

OECD. 2012. OECD Guidelines for Testing of Chemicals. France: OECD. Details available at http://www.oecd.org/env/ehs/testing/oecdguidelinesforthetestingofchemicals.htm#Test_Guidelines, last accessed on 4 April 2012

Szmedra, P. 1999. The health impacts of pesticide use on sugarcane farmers in Fiji. *Asia-Pacific Journal of Public Health* 11(2): 82–8

Tharp, C. 2008. Assessing pesticide toxicity. A self-learning resource from MSU extension, Montana State University extension MT 200810 New 9/08, 500-908SA. Details available at http://www.pesticides.montana.edu/Reference/AssessingToxicity.pdf, last accessed on 28 June 2012

USEPA. 1994. Using toxicity tests in ecological risk assessment, Publication 9345.0-05I, ECO update. Details available at http://www.epa.gov/oswer/riskassessment/ecoup/pdf/v2no1.pdf, last accessed on 2 March 2012

USEPA. 2012. Lethal Dosage (LD_{50}) Values. Environmental Protection Agency, United States. Details available at http://www.epa.gov/oecaagct/ag101/pestlethal.html., last accessed on 5 November 2012

Whitford, F., C. R. Edwards, J. J. Neal, A. G. Martin, J. Osmun, and R. M. Hollingworth. 1992. *Pesticides and Personal Safety,* A. Blessings (ed.), Vol. 20. West Lafayette, Indiana: Purdue pesticide Programs, Purdue University Cooperative Extension Services. Details available at http://pulse.pharmacy.arizona.edu/resources/fertilizers/purdue_pesticide_program.pdf, last accessed on 24 May 2012

WHO. 1997. *Guidelines for Predicting Dietary Intake of Pesticide Residues , revised edition, GEMS/ Food Document.* Switzerland: Programme of food safety and food aid, WHO. Details available at http://www.who.int/foodsafety/publications/chem/en/pesticide_en.pdf, last accessed on 25 November 2013

WHO. 2011. The Joint FAO/WHO Meeting on Pesticide Residues (JMPR). Details available at http://www.fao.org/fileadmin/templates/agphome/documents/Pests_Pesticides/JMPR/crit_review.pdf, last accessed on 25 April 2014

CHAPTER 3

TOXICITY PROFILE

The toxicity profile or toxicological profile of a substance is indicative of the harmful nature of the substance. It is determined on the basis of the different levels of exposure of the substance resulting from different routes of entry.[1]

At present, the use of pesticide is more of a necessity rather than an option. If pesticides are not used against vectors of diseases, then the result will be devastating both for human and animal health and for the economy of the country. Despite several benefits, pesticides have a major impact on the environment and human and animal health. In this chapter, we discuss the toxicity profiles of some substances.

ORGANOPHOSPHATES

The term "organophosphate" is generally understood as an organic derivative of phosphoric acid or similar acid. In pesticide toxicology, it means organophosphorus anticholinesterases. It does not include glyphosate as it does not have insecticidal action. Organophosphates are used on arable crops, in horticulture, and as ectoparasiticides.

On the basis of their toxic action on insects, organophosphate insecticides are grouped into different categories. Examples of organophosphate pesticides that act similar to chlorinated hydrocarbons and as contact poisons include malathion, paraoxon, parathion, and potasan. Selective systemic organophosphate insecticides are absorbed into the plant sap and remain active for long periods of time. These organophosphates are toxic to plant pests but not to their predators.[2]

---

[1] Details available at <www.businessdictionary.com/definition/toxicity-profile.html>; last accessed on 3 March 2011

[2] Details available at <www.michigan.gov/dnr/0,4570,7-153-10370_12150_12220-27249-,00.html>, last accessed on 3 March 2011

Examples of selective systemic organophosphate insecticides are dimefox, mipafox, and schradan.

Organophosphate compounds inhibit acetylcholinesterase enzyme at cholinergic junctions of the central, peripheral, and autonomic nervous systems. These junctions include postganglionic parasympathetic neuroeffector junctions (sites of muscarinic activity), autonomic ganglia and the neuromuscular junctions (sites of nicotinic activity), and certain synapses in the central nervous system. Acetylcholine is the neurotransmitter at cholinergic junctions. Enzyme acetylcholinesterase terminates the action of the neurotransmitter acetylcholine following stimulation of a nerve. The inhibition of acetylcholinesterase makes acetylcholine to accumulate, which results in initial excessive stimulation followed by depression.[3]

Organophosphate compounds have varying toxic capabilities. In comparison to other types of insecticides, organophosphate insecticides produce little or no tissue residues. However, all of them have a cumulative effect with chronic exposure resulting in progressive inhibition of cholinesterase. Organophosphates in liquid form can be absorbed by all routes. However, malathion, the least toxic of all organophosphate compounds, is only slightly absorbed through skin.[4]

Organophosphate insecticides, such as diazinon, chlorpyrifos, disulphoton, azinphos-methyl, and fonofos, have been used widely in agriculture as pesticides. They inhibit acetycholinesterase, thereby affecting neuromuscular transmission (Dyro and Berman 2012). Mammalian toxicities for organophosphate pesticides are given in Table 1.

Chlorpyrifos

Of all the organophosphate pesticides, chlorpyrifos is the most rampantly used. It is a crystalline organophosphate insecticide with IUPAC name O,O-diethyl O-3,5,6-trichloropyridin-2-yl phosphorothioate. As other organophosphate compounds, chlorpyrifos inhibits acetylcholinesterase enzyme, thus acting on the nervous system. This pesticide has been found to be moderately toxic to humans. Chlorpyrifos exposure leads to neurological effects, development disorders, and autoimmune disorders. The acute toxicities in mammal produced by chlorpyrifos are given in Table 2.

The 4 h inhalation LC_{50} dose for chlorpyrifos in rats is greater than 0.2 mg/L (Dow Chemical Company 1986).

---

[3] Details available at <www.michigan.gov/dnr/0,4570,7-153-10370_12150_12220-27249--,00.html>, last accessed on 3 March 2011

[4] Details available at <www.michigan.gov/dnr/0,4570,7-153-10370_12150_12220-27249--,00.html>, last accessed on 3 March 2011

Table 1 Organophosphate mammalian toxicities (mg/kg of body weight)

Common name	Rat oral LD_{50}	Rabbit dermal LD_{50}
Acephate	1,030–1,447	>10,250
Azinphos-methyl	4	150–200 (rat)
Chlorpyrifos	96–270	2,000
Diazinon	1,250	2,020
Dimethoate	235	400
Disulphoton	2–12	3.6–15.9
Ethoprop	61.5	2.4
Fenamiphos	10.6–24.8	71.5–75.7
Malathion	5,500	>2,000
Methamidophos	13 (female only)	122
Methidathion	25–44	200
Methyl parathion	6	45
Naled	191	360
Oxydemeton-methyl	50	1,350
Phorate	2–4	20–30 (guinea pig)
Phosmet	147–316	>4,640
Profenofos	358	472

Source   Fishel (2013a)

Table 2 Mammalian acute toxicity of chlorpyrifos (mg/kg of body weight)

Mammal	Dosage	Reference
	Oral LD_{50}	
Rats	95–270	Gallo and Lawryk 1991; Kidd and James 1991
Rabbits	1,000	Gallo and Lawryk 1991; Kidd and James 1991; Gosselin, Smith, and Hodge 1984
Chicken	32	Gallo and Lawryk 1991; Kidd and James 1991; Gosselin, Smith, and Hodge 1984
Sheep	80	Gallo and Lawryk 1991; Kidd and James 1991; Gosselin, Smith, and Hodge 1984
Guinea pigs	500–504	Gallo and Lawryk 1991; Kidd and James 1991; Gosselin, Smith, and Hodge 1984
Mice	60	Gallo and Lawryk 1991; Kidd and James 1991; Gosselin, Smith, and Hodge 1984
	Dermal LD_{50}	
Rabbits	1,000–2,000	Gallo and Lawryk 1991; Kidd and James 1991
Rats	>2,000	Gallo and Lawryk 1991; Kidd and James 1991

Chronic toxicity

Dogs fed technical chlorpyrifos for 2 years showed 3.0 mg/kg/day increase in liver weight. Signs of cholinesterase inhibition occurred at 1 mg/kg/day. When rats and mice were fed technical chlorpyrifos for 104 weeks, apart from cholinesterase inhibition no other adverse effects were observed (USEPA 1989d). Moderate depression of cholinesterase was observed in rats when they were fed 1 and 3 mg/kg/day dose of chlorpyrifos for a 2-year period. When the experiment came to an end, the cholinesterase levels returned to normal (Gallo and Lawryk 1991). Similar results were obtained when this experiment was conducted in dogs. Neither the dogs nor rats showed any long-term health effects (ACGIH 1986; Gallo and Lawryk 1991).

Dichlorvos

Dichlorvos is an organophosphate insecticide. It is used to control household, public health, and stored product insects. The IUPAC name is 2, 2-dichlorovinyl dimethyl phosphate. Effective against mushroom flies, aphids, spider mites, caterpillars, thrips, and whiteflies, it is also used against parasitic worm infections in dogs, livestock, and humans. As other organophosphates, it also affects acetylcholinesterase thus disrupting the nervous system. The acute toxicity of dichlorvos in different mammals is given in Table 3.

Table 3 Mammalian acute toxicity in Dichlorvos (mg/kg of body weight)

Mammal	Dosage	Reference
Oral LD_{50}		
Rats	25–80	Gallo and Lawryk 1991; Kidd and James 1991; NLM 1995
Rabbits	11–12.5	Gallo and Lawryk 1991; Kidd and James 1991; NLM 1995
Pigs	157	Gallo and Lawryk 1991; Kidd and James 1991; NLM 1995
Dogs	100–1,090	Gallo and Lawryk 1991; Kidd and James 1991; NLM 1995
Chickens	15	Gallo and Lawryk 1991; Kidd and James 1991; NLM 1995
Mice	61–175	Gallo and Lawryk 1991; Kidd and James 1991; NLM 1995
Dermal LD_{50}		
Mice	206	Gallo and Lawryk 1991; Kidd and James 1991; NLM 1995
Rabbits	107	Gallo and Lawryk 1991; Kidd and James 1991; NLM 1995
Rats	70.4–250	Gallo and Lawryk 1991; Kidd and James 1991; NLM 1995

The 4 h LC_{50} for dichlorvos is greater than 0.2 mg/L in rats (NLM 1995).

Chronic toxicity

The cholinesterase activity in the plasma, red blood cells, and brain of rats decreased substantially when they were exposed to air concentrations of 0.5 mg/L dichlorvos for a 5-week period. In a 2-year feeding study, dogs were given 1.6 or 12.5 mg/kg/day of dichlorvos dietary doses. It was observed that their red blood cell cholinesterase activity decreased and liver weight and liver cell size increased (USEPA 1988a). The dietary fluid in the lungs build up when the animals were chronically exposed to dichlorvos.

Fenthion

Fenthion is an organophosphate insecticide used against sucking and biting insects. As other organophosphate pesticides, it inhibits the function of cholinesterase. Its IUPAC name is O, O-dimethyl O-[3-methyl-4-(methylsulfanyl) phenyl] phosphorothioate. Apart from being used against insects, it is also used to control birds, particularly weaver birds, as it is highly toxic to birds.

Fenthion is moderately toxic via the oral and dermal routes, and only slightly toxic via inhalation. The acute toxic effects of fenthion are the same as any other organophosphates but they may take longer period to develop (Gallo and Lawryk 1991). Fenthion is of sufficiently low toxicity; it has been investigated as an agent against insect parasites in animals (such as dogs) (Gallo and Lawryk 1991). In rats, 1 h airborne LC_{50} for fenthion is 2.4–3.0 mg/L (Kidd and James 1991). Table 4 gives the acute toxicity of fenthion on different mammals.

Chronic toxicity

When rats where fed 12.5 mg/kg/day of fenthion, within four weeks weight loss and 85% inhibition of normal brain cholinesterase activity were observed (Gallo and Lawryk 1991). At a dose of 2.5 mg/kg/day, less

Table 4 Mammalian acute toxicity of fenthion (mg/kg of body weight)

Mammal	Dosage	Reference
Oral LD_{50}		
Rats	180–298	Kidd and James 1991; NLM 1995
Rabbits	150	Kidd and James 1991; NLM 1995
Mice	88–145	Kidd and James 1991; NLM 1995
Dermal LD_{50}		
Mice	500	Kidd and James 1991; NLM 1995
Rats	330–1,000	Kidd and James 1991; NLM 1995

severe but still detectable decreases were noticeable (Gallo and Lawryk 1991). Repeated or prolonged exposure to organophosphates may result in the same effects as acute exposure, including delayed symptoms. Dogs fed 1.25 mg/kg/day of fenthion for 1 year showed neither weight loss nor decreased food consumption (Kidd and James 1991).

Malathion

Malathion is a low human toxicity pesticide and it was one of the earliest pesticides developed. Malathion is used against mosquitoes, flies, household pests, and animal parasites. The IUPAC name is diethyl 2-[(dimethoxyphosphorothioyl) sulfanyl] butanedioate. It binds irreversibly to cholinesterase.

Malathion is slightly toxic via both oral and dermal routes. Effects of malathion are similar to those observed with other organophosphates, except that larger doses are required to produce them (Gallo and Lawryk 1991; NLM 1995). Reports suggest that single doses of malathion are likely to affect immune system response (Gallo and Lawryk 1991). Table 5 gives the acute toxicity of malathion on different mammals.

Chronic toxicity

In a study conducted on humans, volunteers were fed very low doses of malathion for 11–12 months. They showed no substantial effects on blood cholinesterase activity. For a period of over 2 years, rats were fed a dietary dose of 5–25 mg/kg/day of malathion. Apart from the depressed cholinesterase activity, no other symptoms were observed in them. When for eight weeks rats were fed small amounts of malathion, no adverse effects were observed in their whole-blood cholinesterase activity (Gallo and Lawryk 1991). Weaning male rats were twice as susceptible to malathion as adults.

Table 5 Mammalian acute toxicity of malathion (mg/kg of body weight)

Mammal	Dosage	Reference
\multicolumn{3}{c}{Oral LD$_{50}$}
Rats	1,000–10,000	Gallo and Lawryk 1991; Kidd and James 1991
Mice	400–4,000	Gallo and Lawryk 1991; Kidd and James 1991
\multicolumn{3}{c}{Dermal LD$_{50}$}
Rats	4,000	Gallo and Lawryk 1991; Kidd and James 1991

Methyl Parathion

Methyl parathion is a potent organophosphate pesticide and it is used to control sucking pests of agricultural crops. Its IUPAC name is *O, O*-diethyl *O*-(4-nitrophenyl) phosphorothioate. Methyl parathion inhibits acetylcholinesterase, thus disrupting the nervous system. It is highly toxic to non-target animals and humans.

Methyl parathion is highly toxic via both oral routes and dermal routes. Table 6 gives the acute toxicity of methyl parathion. The 1 h inhalation LC_{50} for methyl parathion in rats is 0.24 mg/L (Gallo and Lawryk 1991).

Chronic toxicity

Dogs fed 1.25 mg/kg of methyl parathion for 12 weeks showed a significant depression of red blood cells and plasma cholinesterase. However, 0.125 mg/kg of methyl parathion did not lead to any effects (Gallo and Lawryk 1991). Among human volunteers, doses of 1–22 mg/person/day did not show any effect on cholinesterase activity. When volunteers were given dosages of 22, 24, 26, 28, or 30 mg/person/day for four weeks, mild cholinesterase inhibition was observed in some individuals in the 24, 26, and 28 mg dosage groups. In the 30 mg/person/day (about 0.43 mg/kg/day) group, red blood cholinesterase activity was depressed by 37%.

ORGANOCHLORINE PESTICIDES

Organochlorine pesticides are used in agriculture and mosquito control. They are chlorinated hydrocarbons and compounds in this group, include DDT, methoxychlor, dieldrin, chlordane, toxaphene, mirex, kepone, lindane,

Table 6 Mammalian acute toxicity of methyl parathion (mg/kg of body weight)

Mammal	Dosage	Reference
	Oral LD_{50}	
Rats	6–50	Gallo and Lawryk 1991; Kidd and James 1991
Mice	14.5–19.5	Gallo and Lawryk 1991; Kidd and James 1991
Rabbits	420	Gallo and Lawryk 1991; Kidd and James 1991
Guinea pigs	1,270	Gallo and Lawryk 1991; Kidd and James 1991
Dogs	90	Gallo and Lawryk 1991; Kidd and James 1991
	Dermal LD_{50}	
Mice	1,200	Gallo and Lawryk 1991; Kidd and James 1991
Rabbits	300	Gallo and Lawryk 1991; Kidd and James 1991
Rats	67	Gallo and Lawryk 1991; Kidd and James 1991

and benzene hexachloride.[5] Mammalian toxicities for organochlorine pesticides are shown in Table 7.

Chlordane

Chlordane is a polycyclic chlorinated hydrocarbon class of pesticides. The chemical formula is 1, 2, 4, 5, 6, 7, 8, 8-octachloro-3a, 4, 7, 7a-tetrahydro-4, 7-methanoindane. It is a broad-spectrum insecticide and used to control pests in agricultural crops, such as corn and citrus, on lawns and domestic gardens and as a termiticide. Chlordane bioaccumulates and is highly toxic to aquatic life. It may be carcinogenic to humans. The acute toxicity of chlordane in different mammals is given in Table 8. The 4 h inhalation LD_{50} in cats is 100 mg/L (NLM 1995).

Chronic toxicity

When animals were fed chlordane, their liver and the central nervous system were found to be damaged (ATSDR 1989b; USEPA 1989a). When rats were fed a near-lethal dose of 300 mg/kg/day in a 2-year feeding study, haemorrhage was produced in their eyes and nose and changes were observed in their, liver, kidney, heart, lungs, adrenal gland, and spleen tissues. However, rats fed 5 mg/kg/day showed no adverse effects. When mice were fed 22–63.8 mg/kg/day of chlordane for a long period, they lost weight, their liver weight increased, and they died. Dogs fed doses of 15–30 mg/kg/day showed increased liver weights (ATSDR 1989b; Smith 1991).

Dicofol

Dicofol is an organochlorine acaricide. It is a highly toxic pesticide and its chemical formulation is similar to DDT. It is not soluble in water. Dicofol has harmful effects on aquatic animals, and birds exposed to dicofol lay eggs that have thin shells. World Health Organization (WHO) has classified dicofol as moderately hazardous. Humans can be exposed

Table 7 Mammalian toxicity of chlorinated hydrocarbon pesticides (mg/kg of body weight)

Common name	Rat oral LD_{50}	Rabbit dermal LD_{50}
Dicofol	570–595	>2,000
Endosulfan	18–220	200–359

Source Fishel (2013b)

---

[5] Details available at <dhss.delaware.gov/dph/files/organochlorpestfaq.pdf>, last accessed on 1 April 2011

Table 8 Acute mammalian chlordane toxicities (mg/kg of body weight)

Mammal	Dosage	Reference
Oral LD$_{50}$		
Rats	200–700	Kidd and James 1991; Smith 1991
Mice	145–430	Kidd and James 1991; Smith 1991
Rabbits	20–300	Kidd and James 1991; Smith 1991
Hamsters	1,720	Kidd and James 1991; Smith 1991
Dermal LD$_{50}$		
Rabbits	780	Kidd and James 1991; NLM 1995
Rats	530–690	Kidd and James 1991; NLM 1995

to dicofol through ingestion, inhalation, dermal contact, and eyes. In the central nervous system, it inhibits certain enzymes. It causes nausea, dizziness, weakness, vomiting, rash, and conjunctivitis. In severe cases, it results in convulsions, coma, or death from respiratory failure. Table 9 gives the acute mammalian dicofol toxicity (Kamrin 2000).

The inhalation LC$_{50}$ (4 h) in rats is greater than 5 mg/L (Edwards, Ferry, and Temple 1991; Kidd and James 1991; Rohm and Hass Company 1991).

Chronic toxicity

For three months, 12 dogs were fed 25 mg/kg/day of dicofol. Only two dogs were found to survive. At 7.5 mg/kg/day dose, poisoning symptoms and effects on the liver, heart, and testes were observed (Hurt 1991). Dicofol affected the liver of dogs when they were fed a dose of 4.5 mg/kg/day for 1 year. Rats dermally exposed to dicofol as an emulsifiable concentrate

Table 9 Acute mammalian dicofol toxicities (mg/kg of body weight)

Mammal	Dosage
Oral LD$_{50}$	
Rats	575–960
Guinea pigs	1,810
Rabbits	1,810
Mice	420–675
Dermal LD$_{50}$	
Rabbits	2,000–5,000
Rats	1,000–5,000

formulation for a long period of time showed toxic effects on the liver (Hurt 1991). When rats were fed 2.5 mg/kg/day and above dose of dicofol for 2 years, liver growth, enzyme induction, and other changes in the liver, adrenal gland, and urinary bladder resulted. In mice, a dose of 6.25 mg/kg/day and above for three months showed effects on the liver, kidney, and adrenals, and also led to reduced body weights (Rohm and Hass Company 1991).

Endosulfan

Endosulfan is an organochlorine insecticide and acaricide. WHO has classified it as a moderately toxic pesticide. It is used on a variety of vegetables and fruits, on cotton, and on ornamental plants. Endosulfan is volatile, persistent, and bioaccumulative. It is toxic to aquatic organisms. Endosulfan exposure takes place through inhalation, skin, eyes, and ingestion. It causes headache, dizziness, nausea, vomiting, incoordination, tremor, mental confusion, and hyperexcitable state. In extreme cases, it leads to seizure, convulsions, respiratory problems, depression, and coma. Table 10 gives the acute mammalian endosulfan toxicity.

Endosulfan may be slightly toxic via inhalation, with a reported inhalation LC_{50} of 21 mg/L for 1 h and 8.0 mg/L for 4 h (Smith 1991). It is not found to cause skin or eye irritation in animals (Smith 1991).

Chronic toxicity

A high rate of mortality was observed in rats when they were fed an oral dose of 10 mg/kg/day for 15 days. Liver enlargement and other effects were observed when the dose was reduced to 5 mg/kg/day for the same period (Smith 1991). This dose level also caused seizures commencing 25–30 min after dose administration that persisted for approximately 60 min (Smith 1991). When rats were given this dose for 2 years, a reduction in growth

Table 10 Mammalian acute toxicity of endosulfan (mg/kg of body weight)

Mammal	Dosage	Reference
Oral LD_{50}		
Rats	18–160	Kidd and James 1991; Smith 1991
Mice	7.36	Kidd and James 1991; Smith 1991
Dogs	77	Kidd and James 1991; Smith 1991
Dermal LD_{50}		
Rats	78–359	Kidd and James 1991; Smith 1991

and survival, changes in kidney structure, and changes in blood chemistry were observed (ATSDR 1990; Smith 1991).

Heptachlor

Heptachlor is a man-made chemical which is white in colour. Its other trade names are Heptagram, Basaklor, Velsiol 104, and so on. It is highly toxic to aquatic organisms and bird species. Humans get exposed to heptachlor via dermal, inhalation, and ingestion. Humans exposed to the insecticide may show signs of irritability, salivation, lethargy, dizziness, laboured respiration, muscle tremors, and convulsions. Table 11 gives the acute mammalian heptachlor toxicity. Heptachlor is high to moderately toxic orally and it is moderately toxic dermally.

Chronic toxicity

Exposure of heptachlor irrespective of whether it is acute or chronic causes the same effects. Rats given doses of 0.25 mg/kg/day for over 2 years showed no effects (ATSDR 1989a). The mortality rate in mice increased when 1.5 mg/kg/day of heptachlor/heptachlor epoxide was fed to them for 2 years (ATSDR 1989a; Smith 1991). No serious effects were noticed in dogs when they were fed 0.1 mg/kg/day for 60 days. Liver damage was observed when rats were fed a dose of 0.35 mg/kg/day for 50 weeks (ATSDR 1989a). Liver function returned to normal after 30 weeks following discontinuation of dose (Smith 1991).

Lindane

Lindane is an organochlorine pesticide and it is also called gamma-HCH. This is because it comprises 99% of the gamma-isomer of hexachlorocyclohexane

Table 11 Mammalian acute toxicity of heptachlor (mg/kg of body weight)

Mammal	Dosage	Reference
	Oral LD_{50}	
Rats	100–220	ATSDR 1989a; Smith 1991
Mice	30–68	ATSDR 1989a; Smith 1991
Hamsters	100	ATSDR 1989a; Smith 1991
Chicken	62	ATSDR 1989a; Smith 1991
Guinea pigs	116	ATSDR 1989a; Smith 1991
	Dermal LD_{50}	
Rabbits	>2,000	Kidd and James 1991
Rats	119–320	Kidd and James 1991

(HCH). WHO has classified lindane as moderately hazardous. It is used in crops and seed treatment, while in humans it is used for treating lice and scabies. When humans get exposed to lindane, they may suffer from headaches, nausea, dizziness, tremors, and muscular weakness.

Acute toxicity

Lindane is a moderately toxic compound via oral and dermal routes. It is also known as gamma-hexachlorocyclohexane and considered the most acutely toxic of the isomers following single administration (Smith 1991). Lindane is a skin and eye irritant (Kidd and James 1991). It is reported to affect younger animals (Smith 1991). Dermal application of lindane on calves results in toxic effects (Smith 1991). Table 12 gives the acute mammalian lindane toxicity.

Notably, a 1% solution of lindane in vanishing crème resulted in a sixfold increase in acute toxicity via the dermal route in rabbits, with a reported dermal LD_{50} of 50 mg/kg (Smith 1991).

Chronic toxicity

When dogs, rats, and mice were administered a dose of 1.25 mg/kg/day for 2 years, no adverse effects were observed in them (Matsumura 1985; Smith 1991). When 40–80 mg/kg/day of lindane was given to dogs for 2 years, it caused fatal effects on them. An increased dose of 2.6–5.0 mg/kg/day led to convulsions in some test animals (Smith 1991) and liver lesions in rats (Smith 1991). In a 2-year study conducted on rats, a dose of 5 mg/kg/day of lindane caused significant liver changes (Smith 1991). When the dose of 6–10 mg/kg/day was given to mice in two different studies, it resulted in different outcomes. In one study, the mice were not

Table 12 Mammalian acute toxicity of lindane (mg/kg of body weight)

Mammal	Dosage	Reference
Oral LD_{50}		
Rats	88–190	Smith 1991
Mice	59–562	Kidd and James 1991; Smith 1991
Guinea pigs	100–127	Kidd and James 1991; Smith 1991
Rabbits	200	Kidd and James 1991; Smith 1991
Dermal LD_{50}		
Mice	300	Kidd and James 1991; Smith 1991
Guinea pigs	400	Kidd and James 1991; Smith 1991
Rabbits	300	Kidd and James 1991; Smith 1991
Rats	500–1,000	Kidd and James 1991; Smith 1991

affected while in the other study the dose caused metabolic changes in liver (Smith 1991). Higher doses of lindane led to liver damage in mice. It also caused damage to kidney, pancreas, testes, and nasal mucous membrane in test animals (Smith 1991).

CARBAMATES

Carbamates are organic compounds. They are derived from carbamic acid. As organophosphate insecticides, carbamate insecticides also inhibit cholinesterase enzymes. However, unlike organophosphates, their action on cholinesterase is brief.[6] Carbamates are used to kill pests in homes, gardens, and agriculture. Carbamate pesticides differ in their spectrum of activity, mammalian toxicity, and persistence. They are relatively unstable compounds that break down in the environment within weeks or months. Mammalian toxicities for carbamate pesticides are shown in Table 13.

Acute toxicity

Onset and severity of symptoms are dose related in the case of carbamate pesticides. Excessive salivation and heavy breathing are observed after 30 min following administration of a low dose of carbamate pesticide. When the dose level is higher, it will cause excessive tearing, urination, uncontrollable defecation, nausea, and vomiting (Baron 1991). At the maximum dose, additional symptoms develop that include violent intestinal movements, muscle spasms, and convulsions. In some cases death occurs owing to respiratory failure resulting from paralysis of respiratory muscles (Baron 1991).

Table 13 Carbamate pesticide mammalian toxicities (mg/kg of body weight)

Common name	Rat oral LD_{50}	Rabbit dermal LD_{50}
Aldicarb	1	20
Carbaryl	500–850	>2,000
Carbofuran	8	>3,000
Fenoxycarb	16,800	>2,000
Methiocarb	60–1,000 depending on product	>2,000 (rat)
Methomyl	30–34	>2,000
Oxamyl	5.4	2,960
Thiodicarb	66	>2,000

Source <www.health.qld.gov.au/ph/documents/ehu/4174.pdf>, last accessed on 3 March 2011

---

[6]Details available at <www.health.qld.gov.au/ph/documents/ehu/4174.pdf>, last accessed on 20 April 2011

The LD_{50} of aldicarb, when given in the liquid or oil form, in rats, mice, guinea pigs, and rabbits ranges between 0.5 mg/kg and 1.5 mg/kg. The LD_{50} of dry aldicarb granules is 7.0 mg/kg. The toxicity is lower, in this case but it is still on the higher side (Baron 1991; USEPA 1987b). The oral LD_{50} for bendiocarb is 34–156 mg/kg in rats, 35–40 mg/kg in rabbits, and 35 mg/kg in guinea pigs. The dermal LD_{50} in rats is 566 mg/kg (Baron 1991). The 4 h acute inhalation LC_{50} in rats is 0.55 mg/L air (Kidd and James 1991).

The LD_{50} in rats, when administered carbaryl orally, ranges between 250 mg/kg and 850 mg/kg. In mice, it ranges between 100 mg/kg and 650 mg/kg (NLM 1995; USEPA 1987b). The inhalation LC_{50} in rats is greater than 200 mg/L (Kuna and Heal 1948). In rabbits, it was observed that when they were given low doses of carbaryl, it led to minor skin and eye irritation. The dermal LD_{50} of carbaryl in rabbits is measured greater than 2000 mg/kg (NLM 1995). The oral LD_{50} is 5–13 mg/kg in rats, 2 mg/kg in mice, and 19 mg/kg in dogs. The dermal LD_{50} is greater than 1000 mg/kg in rabbits (Baron 1991). The 4 h LC_{50} for inhalation of carbofuran is 0.043–0.053 mg/L in guinea pigs (Kidd and James 1991).

Chlorpropham shows a low level of toxicity profile and no acute toxicity is observed after administration of less than 1000 mg/kg/day of chlorpropham. In rats, the oral chlorpropham LD_{50} is between 5000 mg/kg and 7500 mg/kg (Kidd and James 1991), while it is 5000 mg/kg in rabbits (Hallenbeck and Cunningham-Burns 1985). The 4 h inhalation LC_{50} in rats for chlorpropham is greater than 32 mg/L. Mammals exposed to fenoxycarb orally were not found to experience any harmful effects. For rats exposed to fenoxycarb orally, the LD_{50} is greater than 10,000 mg/kg (Kidd and James 1991). Through the dermal route also the compound does not cause any toxic effect. In rats, the dermal LD_{50} is greater than 2000 mg/kg (Kidd and James 1991).

Methomyl and oxamyl are two highly toxic compounds via the oral route. The oral LD_{50} of methomyl in rats is 17–24 mg/kg (Kidd and James 1991), in mice it is 10 mg/kg, and in guinea pigs 15 mg/kg (Baron 1991). For oxamyl, the oral LD_{50} in rats is 5.4 mg/kg (Hallenbeck and Cunningham-Burns 1985). Similar to other carbamate pesticides, oxamyl inhibits the cholinesterase enzyme, which is short term and reversible, and leads to effects such as headache, nausea, sweating, tearing, tremors, and blurred vision. (Baron 1991). Dermal exposure to oxamyl is slightly toxic. The dermal LD_{50} in rabbits for technical oxamyl is 2960 mg/kg (Kidd and James 1991). Propoxur, another carbamate pesticide, is highly toxic via the oral route and only slightly toxic via the dermal route. Oral

LD_{50} value in rats and mice is 100 mg/kg (Kidd and James 1991) and it is 40 mg/kg in guinea pigs (Baron 1991). The dermal LD_{50} value in rats is from 1000 mg/kg to greater than 2400 mg/kg, and it is greater than 500 mg/kg in rabbits (Baron 1991). Tests show that propoxur does not cause skin or eye irritation in rabbits (Kidd and James 1991). It is also slightly toxic, via the inhalation route, with a reported 1 h inhalation LC_{50} of greater than 1.44 mg/L (Kidd and James 1991).

Chronic toxicity

Unlike organophosphorus compounds, carbamate compounds do not cause a delayed neurotoxic reaction when inhibiting cholinesterase. A 22-day experiment was conducted on rats to study neurotoxicity. These rats were given various acute doses of carbamate compounds that were equivalent to half and up to several times the LD_{50} values for them. No behavioural neurotoxicity was observed in the rats (Baron 1991; Kaloyanova and El Batawi 1991).

Aldicarb exposure does not lead to chronic effects. For 2 years, rats and dogs were fed low doses of aldicarb and they did not show any significant adverse effects (USEPA 1987b). One epidemiological study suggested a possible link between low-level exposure and immunological abnormalities (USEPA 1987b). However, the results of this study were widely disputed. In another study, rats were administered a high dose of 10 mg/kg/day of bendiocarb for 2 years. It was observed that there was considerable changes in the organ weights, blood, and urine characteristics. Incidences of stomach and eye lesions also considerably increased (Baron 1991).

Rats fed high doses of carbaryl for long time did not show any symptoms of reproductive or foetal effects (NLM 1995). On the other hand, rats fed very high doses of carbofuran (5 mg/kg/day) for 2 years showed decrease in weight. Same results were obtained in mice when a similar test was conducted on them (Baron 1991). Prolonged or repeated exposure to carbofuran may cause the same effects as an acute exposure (Baron 1991).

When rats were given fenoxycarb at a low dose of 10 mg/kg/day or less for 1 year, no side effects were seen. Dogs fed the compound at 15.9 mg/kg/day or less for 11–12 months also did not experience any side effects (USEPA 1987b). In another study conducted for 2 years, rats were fed 2.5, 5, or 20 mg/kg/day. Side effects were visible only at the dose of 20 mg/kg/day, in which case female rats showed reduced red blood cell counts and haemoglobin levels (USEPA 1987b). It is unlikely that chronic effects would be observed in humans unless they are highly exposed, such as in the case of chronic misuse.

In the case of oxamyl, symptoms for repeated exposure will be the same as that of acute exposure. In a 2-year mouse feeding study, no effects were observed at a dose of 1.25 mg/kg/day. However, at the very high doses of 2.5–3.75 mg/kg/day, decreased body weight and changed nutritional performance were observed (DuPont de Nemours 1984). Prolonged or repeated exposure to propoxur may cause symptoms similar to acute effects. Propoxur is effectively detoxified (transformed into less toxic or practically non-toxic forms), thus enabling rats to tolerate daily doses approximately equal to the LD_{50} of the insecticide for long periods, provided the dose is spread out over the entire day rather than ingested all at once (Baron 1991).

PYRETHROIDS AND OTHER BOTANICAL PESTICIDES

On the basis of biological origins, pyrethroids and other botanical pesticides are grouped together (Kamrin 2000). Pesticides have two types of origins. There are pesticides which are extracted from naturally occurring materials while others are synthesized in laboratories. Naturally occurring materials are derived from plants. Over long periods of time, many plants have evolved substances to defend themselves against plant-eating insects (Kamrin 2000). The plants use these compounds for their protection from insects. A number of compounds from plants that are effective insecticides or fungicides. Development of these botanical pesticides is one of the fastest growing areas in the pesticide industry (Kamrin 2000).

Acute Toxicity

Pyrethroids are slightly or moderately toxic to animals. Table 14 gives the acute toxicity level of different pyrethroids.

Chronic Toxicity

Pyrethroids when consumed for a long time can affect the liver (Ray 1991). Resmethrin fed to rats at a dose of 125 mg/kg/day in a chronic feeding study caused pathological liver changes apart from increased liver weights (USEPA 1988b). The liver got damaged in rats when they were fed a large dose of pyrethrins for over 2 years (Ray 1991). Fenvalerate compound did not cause any changes in the blood or urine in rats when it was fed at a dose of 12.5 mg/kg/day for 2 years (Kidd and James 1991).

Table 14 Mammalian toxicity of pyrethroid pesticide (mg/kg of body weight)

Pyrethroids	Acute toxicity	Oral LD_{50}	Dermal LD_{50}	LC_{50}	Reference
Cypermethrin	Moderately toxic	187–326 in male rats; 150–500 in female rats; 82–779 in mice	1,600 in rats; >2,000 in rabbits		Ray 1991; USEPA 1983; USEPA 1989c
Nicotinoid	Moderately to highly toxic via oral route				
Allethrin		1,100 in male rats; 685 female rats; 480 in mice; 4,290 in rabbits			Kidd and James 1991; USEPA 1988c
D-Allethrin		1,320 in rats	>2,500 in rats		Kidd and James 1991; WHO 1989
Esfenvalerate	Slightly toxic via dermal route and non-toxic via inhalation	458 in rats	2,500 in rabbits	>2.93 mg/L	Kidd and James 1991; Ray 1991
Technical flucythrinate	Moderately toxic via dermal route	81 in male rats; 67 in female rats; 76 in mice	>1,000 in rabbits; >2,000 in guinea pigs		USEPA 1989c; Kidd and James 1991; Mehler 1989
Fluvalinate	Moderately toxic via oral route	261–281 in rats			Ray 1991; USEPA 1986
Mavrik 2E	Less toxic via oral route	1,050–1,110 in rats			ATSDR 1990
Permethrin	Non-toxic via oral route, slightly toxic via dermal route	430–4,000 in rats	4,000 in rats; >2,000 in rabbits		Kidd and James 1991; Ray 1991

Contd...

Table 14 *Contd...*

Pyrethroids	Acute toxicity	Oral LD$_{50}$	Dermal LD$_{50}$	LC$_{50}$	Reference
Resmethrin	Non-toxic via ingestion and slightly toxic via dermal route and inhalation	>2,500 or >1,244 in rats;	>3,000 in rats; >2,500 in rabbits; >5,000 in mice	>9.49 mg/L	Kidd and James 1991; USEPA 1988b
Rotenone		132–1,500 in rats; 350 in white mice			Kidd and James 1991
Ryania		1,200 and 750 in rats; 2,500 in guinea pigs; 650 in rabbits; 150 in dogs			Soloway 1976; Kuna and Heal 1948; NLM 1995

For 28 days, one group of rats was fed 15 mg/kg/day of flucythrinate while another group was fed 7.5 mg/kg/day. The former group suffered from severe motor symptoms like involuntary muscular movement and the latter one sufferred from moderate motor symptoms. Rats fed a dose of fluvalinate of 3 mg/kg/day for 90 days and dogs fed 5 mg/kg/day for six months did not suffer from any side effects (USEPA 1989d). In other studies, effects observed were increased liver and kidney weights and adverse changes in liver tissues in test animals (USEPA 1986).

Different effects were observed in different animals when they were fed permethrin. Dogs fed 5 mg/kg/day of permethrin for 90 days did not show any adverse effects (Thomson 1985). While a slight increase in liver weight was observed in rats when they were given a dose of 150 mg/kg/day for six months (Hallenbeck and Cunningham-Burns 1985). In chickens, the immune system activity got suppressed when they were fed 0.1 ppm of permethrin for 3–6 weeks after hatching (Hallenbeck and Cunningham-Burns 1985).

In a chronic feeding study with rats, 25 mg/kg/day (the lowest dose tested) caused liver enlargement. At 125 mg/kg/day, pathological liver changes in addition to increased liver weights were observed. Doses of 250 mg/kg/day caused increased thyroid weight and thyroid cysts (USEPA 1989c). In another study conducted for 90 days, doses of 150 mg/kg/day did not lead to any adverse effects in exposed rats (NLM 1995). Increased liver weights were observed in dogs fed with 30 mg/kg/day for 180 days. In this study, no effects were observed in dogs at dose rates of 10 mg/kg/day (USEPA 1989c).

In a five-month study, 1% ryania powder was fed to rats and guinea pigs. This is equal to approximately 600 mg/kg/day for rats and 361 mg/kg/day for guinea pigs. No effects were observed in both the animals (Kuna and Heal 1948). When 5% ryania powder (approximately 2700 mg/kg/day) was fed to rats, extreme weight loss and 100% death were seen within 25 days.

THIOCARBAMATES AND DITHIOCARBAMATES

Carbamate pesticides consist of two subclasses: thiocarbamates and dithiocarbamates. In thiocarbamates, one oxygen atom is substituted for one sulphur atom. In dithiocarbamates, two oxygen atoms are replaced by sulphur (Kamrin 2000).

Most members of thiocarbamate and dithiocarbamate groups of pesticides do not cause any toxic effects via oral, dermal, and inhalation routes. Skin and eye irritation and skin sensitization occur on prolonged application. The acute toxicity LD_{50} of thiocarbamates and dithiocarbamates via oral route ranges from 300 mg/kg to 400 mg/kg to even greater than 5000 mg/kg in rats and other test animals, suggesting they are practically non-toxic (Edwards, Ferry, and Temple 1991; NLM 1995; Stevens and Sumner 1991). In the case of dermal exposure to these two classes of pesticides, the acute LD_{50} values are almost greater than 2000 mg/kg in rats (Dyro and Berman 2012; Fishel 2013b). Although both classes of pesticides are known to be slightly toxic via the inhalation route, the precise LC_{50} values are not known (Edwards, Ferry, and Temple 1991; NLM 1995; Stevens and Sumner 1991; WSSA 1994).

Thiocarbamates

Animals exposed to thiocarbamates for a long time show cholinesterase inhibition, degeneration of nervous tissue (of the spinal cord, muscles, and heart), and increase in liver and thyroid weights (Kamrin 2000). These symptoms were manifested when rats were given a dose of 2 mg/kg/day for two generations and dogs were fed a dose greater than 15 mg/kg/day for over 2 years (Edwards, Ferry, and Temple 1991; Gosselin, Smith, and Hodge 1984; Stevens and Sumner 1991). High variability in some toxic responses was observed across species. The repeated application of some thiocarbamates caused skin irritation (Gosselin, Smith, and Hodge 1984; Stevens and Sumner 1991).

Butylate

Butylate is a member of the thiocarbamate class of chemicals and is a herbicide used in corn to control grassy weeds and some broadleaf weeds. Butylate, in combination with atrazine and cyanazine, is applied to soil immediately before corn is planted. It acts selectively on seeds of weeds that are in the germination stage of development and are absorbed from the soil by shoots of grass seedlings before they emerge, causing slow shoot growth and twisted leaves (Kamrin 2000).

Acute toxicity

Similar to other thiocarbamates, butylate, when exposed through skin and by inhalation, causes skin irritation, scratchy throat, sneezing, and coughing (NLM 1995; WSSA 1994). Slight eye irritation can be caused by butylate, potentially leading to permanent eye damage (USEPA 1983; WSSA 1994). Rabbits exposed to 2000 mg technical butylate (85.71% pure) for 24 h suffered from irritation of skin. The acute dermal LD_{50} for butylate is greater than 4640 mg/kg in rabbits (WSSA 1994; USEPA 1989b).

In male guinea pigs, the oral LD_{50} is 1659 mg/kg and it is 5431 mg/kg in female rats. Butylate's inhalation LC_{50} (2 h) is 19 mg/L (Kidd and James 1991; USEPA 1989b).

Chronic toxicity

In an experiment, rabbits were administered 21 doses of 20 mg/kg/day and 40 mg/kg/day. Apart from local skin irritation, no other effects were observed (USEPA 1989b). Several studies have shown that long-term exposure to high doses of butylates causes increase in liver weights in test animals (USEPA 1989b).

In a separate study, rats were fed 50, 100, 200, or 400 mg/kg/day of butylates for 2 years. It was observed that except for the lowest dose, at all other doses body weights got decreased and liver to body weight ratios increased. In another experiment, rats were fed, for 2 years, 20, 80, or 120 mg/kg/day. Kidney and liver lesions formed at 80 and 120 mg/kg/day doses. Blood clotting was affected in rats fed 10, 30, and 90 mg/kg/day of butylate for 56 weeks. At the last two doses, body weight and testes to body weight ratios decreased, liver to body weight ratios increased, and lesions formed on the testes. Decreased body weights, increased liver weights, and increased incidence of liver lesions were observed in dogs when they were fed 100 mg/kg/day of butylate for 1 year. When 5 mg/ kg/day and 25 mg/kg/day doses of butylate were fed to them for the same period, no adverse affects occurred (USEPA 1989b).

Dithiocarbamates

Long-term exposure to dithiocarbamates leads to symptoms, such as drowsiness, incoordination, weakness, and paralysis. In rats, these symptoms were produced when they were given a dose of 40–50 mg/kg/day. Dogs given 5 mg/kg/day for 1 year showed no adverse effects. Repeated or prolonged exposure to dithiocarbamates can also cause skin sensitization (Edwards, Ferry, and Temple 1991; Stevens and Sumner 1991).

Ethylene(bis)dithiocarbamates

Ethylene(bis)dithiocarbamate (EBDC) pesticides are a class of fungicides used for controlling blights in potatoes and tomatoes and diseases in vegetables, fruits and other crops. It is also used for seed treatment. Chronic exposure to EBDCs leads to the production of ethylenethiourea (ETU). ETU may be produced during metabolism or introduced, as a contaminant during manufacture. It is also generated in small amounts when EBDC residues are present on produce for long periods, or during cooking (Edwards, Ferry, and Temple 1991). ETU has caused goitre and impaired thyroid function, birth defects, and cancers in test animals (Edwards, Ferry, and Temple 1991). There is a possibility that thyroid effects observed in test animals because of exposure to EBDCs for a long time may be due to ETU (Edwards, Ferry, and Temple 1991).

Symptoms, such as thyroid enlargement and impairment were produced when dogs and rats were given mancozeb dose of 2–5 mg/kg/day for two years (Edwards, Ferry, and Temple 1991; USEPA 1987d). In the case of maneb, similar effects were observed in rats when they were fed a dose greater than 12.5 mg/kg/day (Hallenbeck and Cunningham-Burns 1985). At a daily higher dose of EBDCs, gastrointestinal and nervous system disturbances (muscular weakness and tremor) were observed in these animals. No adverse were seen in dogs when they were exposed to 45 mg/kg/day of metiram for 90 days or fed 250 mg/kg/day of zineb for 1 year (Edwards, Ferry, and Temple 1991). Field studies of some EBDCs have shown that they may cause skin or eye irritation and contact dermatitis (Edwards, Ferry, and Temple 1991).

Mancozeb

Mancozeb is used to protect fruits, vegetables, nuts, and field crops from fungal diseases. It is also used for seed treatment of cotton, potatoes, corn, peanuts, and so on. Available in the form of dust, liquid, water dispersible granules, wettable powder, and ready-to-use formulations, mancozeb is used in combination with zineb and maneb.

Acute toxicity

Mancozeb is non-toxic, both orally and dermally. In rats, oral LD_{50} value for mancozeb is greater than 5000 mg/kg to greater than 11,200 mg/kg (Edwards, Ferry, and Temple 1991; Kidd and James 1991). Dermal LD_{50} values is greater than 10,000 mg/kg in rats and greater than 5000 mg/kg in rabbits (NLM 1995). It is a mild skin irritant and sensitizer and a mild to moderate eye irritant in rabbits (DuPont de Nemours 1983; NLM 1995). Workers with occupational exposure to mancozeb have developed sensitization rashes (Edwards, Ferry, and Temple 1991).

As other cholinesterase inhibitors, early symptoms of mancozeb include blurred vision, fatigue, headache, vertigo, nausea, pupil contraction, abdominal cramps, and diarrhoea. Severe inhibition of cholinesterase may cause excessive sweating, tearing, slow heartbeat, giddiness, slurred speech, confusion, excessive fluid in the lungs, convulsions, and coma.

Chronic toxicity

In a long-term study, rats fed dietary doses of 5 mg/kg/day showed no toxicological effects (Edwards, Ferry, and Temple 1991). Doses of 2.5–25 mg/kg/day and 0.625 mg/kg/day were fed to two groups of dogs for 2 years. Dogs fed higher doses showed thyroid problem as a result of lower iodine uptake (Edwards, Ferry, and Temple 1991).

A major toxicological concern in case of chronic exposure is the generation of ETU in the course of mancozeb metabolism, and as a contaminant in mancozeb production (Edwards, Ferry, and Temple 1991; Lu 1995). ETU may also be produced when EBDCs are used on stored produce or during cooking (USEPA 1987d). In addition to having the potential to cause goitre, a condition in which the thyroid gland is enlarged, this metabolite is also found to produce birth defects and cancer in experimental animals (USEPA 1987d).

Maneb

Another EBDC fungicide used for protecting potatoes, tomatoes, fruits, vegetables, field crops, and ornamentals against various diseases is maneb. Apart from protecting against fungal diseases, it is also effective against other diseases and is available as granular, wettable powder, flowable concentrate, and ready-to-use formulations (Kamrin 2000).

Acute toxicity

When ingested orally, maneb is non-toxic. In rats, oral LD_{50} value is greater than 5000–8000 mg/kg and it is greater than

8000 mg/kg in mice (Edwards, Ferry, and Temple 1991; Kidd and James 1991). Maneb is slightly toxic in mice dermally. Dermal LD_{50} of maneb is greater than 5000 mg/kg in rats (Kidd and James 1991). Inflammation or irritation of skin, eyes, and respiratory tract has resulted from contact with maneb (Edwards, Ferry, and Temple 1991; Gosselin, Smith, and Hodge 1984). The 4 h inhalation LC_{50} is greater than 3.8 mg/L, indicating slight toxicity.

Acute exposure to maneb leads to the following effects (Edwards, Ferry, and Temple 1991; Gosselin, Smith, and Hodge 1984):

• Hyperactivity and incoordination
• Loss of muscular tone
• Nausea
• Vomiting
• Diarrhoea
• Loss of appetite
• Weight loss
• Headache
• Confusion
• Drowsiness
• Coma
• Slowed reflexes
• Respiratory paralysis
• Death

Chronic toxicity

When rats were fed 12.5 mg/kg/day for over 2 years, no adverse effects were observed in them (Edwards, Ferry, and Temple 1991). However, a dose of 62.5 mg/kg/day continued for 97 days led to goitre and reduced growth rate (USEPA 1987d).

Tremors, lack of energy, gastrointestinal disturbances, and incoordination were observed in dogs given maneb 200 mg/kg/day orally for three or more months. In addition, spinal cord damage was observed in them (Edwards, Ferry, and Temple 1991). Rats fed 1500 mg/kg/day for 10 days showed weight loss, weakness of hind legs, and increased mortality (Edwards, Ferry, and Temple 1991).

2,4-D

2, 4-Dichlorophenoxyacetic acid (2,4-D) is a herbicide used for controlling broadleaf weeds.

Acute toxicity

The acid form is of slight to moderate toxicity. In rats, oral LD_{50} of 2,4-D ranges from 375 mg/kg to 666 mg/kg, in mice 370 mg/kg, and in guinea pigs from less than 320 mg/kg to 1000 mg/kg. The dermal LD_{50} values are 1500 mg/kg in rats and 1400 mg/kg in rabbits (NLM 1995; Stevens and Sumner 1991; WSSA 1994).

Chronic toxicity

No serious effects were seen in rats given high amounts (50 mg/kg/day) of 2, 4-D in diet for 2 years. Mortality was observed in dogs fed lower amounts of 2,4-D in their food for 2 years. This might be because dogs do not excrete organic acids efficiently. A person administered a total of 16.3 g in 32 days therapeutically lapsed into a stupor and showed signs of incoordination, weak reflexes, and loss of bladder control (NLM 1995; Stevens and Sumner 1991; WSSA 1994).

Screenshots of resource materials for toxicity profile are as follows:

1. Agency for Toxic Substances and Disease Registry

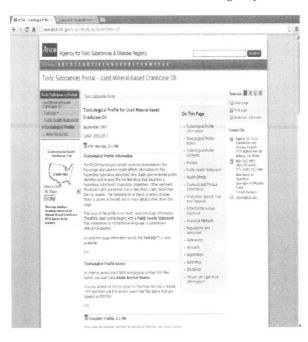

2. University of Florida IFAS extension–pesticide toxicity profile

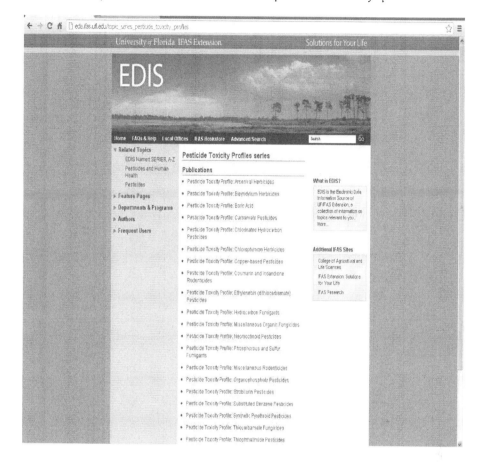

REFERENCES

ACGIH (American Conference of Governmental Industrial Hygienists). 1986. *Documentation of the Threshold Limit Values and Biological Exposure Indices,* 5th edn. Cincinnati, OH: ACGIH

ATSDR (Agency for Toxic Substances and Disease Registry). 1989a. *Toxicological Profile for Heptachlor and Heptachlor Epoxide.* Atlanta, GA: ATSDR

ATSDR (Agency for Toxic Substances and Disease Registry). 1989b. *Toxicological Profile for Chlordane.* Atlanta, GA: ATSDR

ATSDR (Agency for Toxic Substances and Disease Registry). 1990. *Toxicological Profile for Endosulfan,* Draft Report. Atlanta, GA: ATSDR

Baron, R. L. 1991. Carbamate insecticides. In *Handbook of Pesticide Toxicology,* edited by W. J. Hayes, Jr and E. R. Laws, Jr. New York: Academic Press

Dow Chemical Company. 1986. *Summary of Acute Dermal Toxicity Study on Chlorpyrifos in Fischer 344 Rats.* Indianapolis, IN: Dow Chemical Company

DuPont de Nemours, E. I. 1983. *Technical Data Sheet for Mancozeb.* Wilmington, DE: Biochemicals Department

DuPont de Nemours, E. I. 1984. *Technical Data Sheet: Oxamyl.* Wilmington, DE: Agricultural Products Division

Dyro F. M. and F. A. Berman. 2012. Organophosphates. Details available at <http://emedicine.medscape.com/article/1175139-overview>, last accessed on 31 March 2011

Edwards, I. R., D. G. Ferry, and W. A. Temple. 1991. Fungicides and related compounds. In *Handbook of Pesticide Toxicology,* edited by W. J. Hayes, Jr and E. R. Laws, Jr. New York: Academic Press

Fishel, F. M. 2013a. Pesticide toxicity profile: organophosphate pesticides. Details available at <http://edis.ifas.ufl.edu/pi087>, last accessed on 22 January 2011

Fishel, F. M. 2013b. Pesticide toxicity profile: chlorinated hydrocarbon pesticides. Details available at http://edis.ifas.ufl.edu/pi090

Gallo, M. A. and N. J. Lawryk. 1991. Organic phosphorus pesticides. In *Handbook of Pesticide Toxicology,* edited by W. J. Hayes, Jr and E. R. Laws, Jr. New York: Academic Press

Gosselin, R. E., R. P. Smith, and H. C. Hodge. 1984. *Clinical Toxicology of Commercial Products,* 5th edn. Baltimore, MD: Williams and Wilkins

Hallenbeck, W. H. and K. M. Cunningham-Burns. 1985. *Pesticides and Human Health.* New York: Springer-Verlag

Hurt, S. S. 1991. *Dicofol: toxicological evaluation of dicofol prepared for the WHO Expert Group on Pesticide Residues (Report No. 91R-1017).* Spring House, PA: Toxicology Department, Rohm and Haas Company

ITRC (Industrial Toxicology Research Centre). 1990. *Epidemiological Health Survey of Pesticide Sprayers.* Lucknow: ITRC.

Kaloyanova, F. P. and M. A. El Batawi. 1991. *Human Toxicology of Pesticides.* Boca Raton, FL: CRC Press

Kamrin, M. A. 2000. *Pesticide Profiles: toxicity, environmental impact and fate.* New York: Lewis Publishers

Kidd, H. and D. R. James (eds). 1991. *The Agrochemicals Handbook,* 3rd edn. Cambridge: Royal Society of Chemistry Information Services

Kuna, S. and R. E. Heal. 1948. Toxicological and pharmacological studies on the powdered stem of *Ryania speciosa,* a plant insecticide. *Journal of Pharmacology and Experimental Therapeutics* 93: 407–413

Lu, F. C. 1995. A review of the acceptable daily intakes of pesticides assessed by the World Health Organization. *Regulatory Toxicology and Pharmacology* 21: 351–364

Matsumura, F. 1985. *Toxicology of Insecticides,* 2nd edn. New York: Plenum Press

Mehler, L. N. 1989. *Assessment of Human Exposure to Flucythrinate,* HS-1510. Sacramento, CA: California Department of Food and Agriculture

NLM (National Library of Medicine). 1995. *Hazardous Substances Data Bank*. Bethesda, MD: National Library of Medicine

Ray, D. E. 1991. Pesticides derived from plants and other organisms. In *Handbook of Pesticide Toxicology*, edited by W. J. Hayes, Jr and E. R. Laws, Jr. New York: Academic Press

Rohm and Haas Company. 1991. *Material Safety Data Sheet for Kelthane Technical B Miticide*. Philadelphia, PA: Rohm and Haas Company

Singh, V. K., Jyoti, M. M. Reddy, C. Kesavachandran, S. K. Rastogi, and M. K. Siddiqui. 2007. Biomonitoring of organochlorines, glutathione, lipid peroxidation and cholinesterase activity among pesticide sprayers in mango orchards. *Clin Chim Acta*. 377(1-2): 268-72

Smith, A. G. 1991. Chlorinated hydrocarbon insecticides. In *Handbook of Pesticide Toxicology*, edited by W. J. Hayes, Jr and E. R. Laws, Jr. New York: Academic Press

Soloway, S. B. 1976. Naturally occurring insecticides. *Environmental Health Perspectives* 14: 109–116

Stevens, J. T. and D. D. Sumner. 1991. Herbicides. In *Handbook of Pesticide Toxicology*, edited by W. J. Hayes, Jr and E. R. Laws, Jr. New York: Academic Press

Sweet, D. V. 1997. *Registry of Toxic Effects of Chemical Substances*. Cincinnati, OH: US Department of Health and Human Services. Details available at <www.cdc.gov/niosh/pdfs/97-119.pdf>, last accessed on 17 April 2011

Thomson, W. T. 1985. Insecticides. In *Agricultural Chemicals: book I*. Fresno, CA: Thomson Publications

USEPA (US Environmental Protection Agency). 1983. *Guidance for the Reregistration of Pesticide Products Containing Butylate as the Active Ingredient*. Washington, DC: USEPA

USEPA (US Environmental Protection Agency). 1985. Pesticide tolerance on an agricultural commodity: Flucythrinate. *Federal Registry* 50: 21050–21051

USEPA (US Environmental Protection Agency). 1986. *Pesticide Fact Sheet Number 86: fluvalinate*. Washington, DC: Office of Pesticides and Toxic Substances

USEPA (US Environmental Protection Agency). 1987a. *Health Advisories for 50 Pesticides*. Washington, DC: Office of Drinking Water

USEPA (US Environmental Protection Agency). 1987b. *Health Advisory Draft Report: carbaryl*. Washington, DC: Office of Drinking Water

USEPA (US Environmental Protection Agency). 1987c. *Health Advisory Summary: methomyl*. Washington, DC: Office of Drinking Water

USEPA (US Environmental Protection Agency). 1987d. *Pesticide Fact Sheet Number 125: mancozeb*. Washington, DC: Office of Pesticides and Toxic Substances

USEPA (US Environmental Protection Agency). 1988a. Dichlorvos: initiation of special review. *Federal Register* 53: 5542–5549

USEPA (US Environmental Protection Agency). 1988b. *Pesticide Fact Sheet Number 193: resmethrin.* Washington, DC: Office of Pesticides and Toxic Substances

USEPA (US Environmental Protection Agency). 1988c. *Pesticide Fact Sheet Number 158: allethrin stereoisomers.* Washington, DC: Office of Pesticides and Toxic Substances

USEPA (US Environmental Protection Agency). 1988d. *Pesticide Fact Sheet Number 181: metiram.* Washington, DC: Office of Pesticides and Toxic Substances

USEPA (US Environmental Protection Agency). 1989a. *Health Advisory Summary: chlordane.* Washington, DC: Office of Drinking Water

USEPA (US Environmental Protection Agency). 1989b. *Health Advisory Summary: butylate.* Washington, DC: Office of Drinking Water

USEPA (US Environmental Protection Agency). 1989c. *Pesticide Fact Sheet Number 199: cypermethrin.* Washington, DC: Office of Pesticides and Toxic Substances

USEPA (US Environmental Protection Agency). 1989d. *Registration Standard (Second Round Review) for the Reregistration of Pesticide Products Containing Chlorpyrifos.* Washington, DC: Office of Pesticides Program, USEPA

USEPA (US Environmental Protection Agency). 1992. Ethylene bisdithiocarbamates (EBDCs): notice of intent to cancel and conclusion of special review. *Federal Register* 57: 7434–7530

WHO (World Health Organization). 1989. Allethrins: allethrin, D-allethrin, bioallethrin, S-bioallethrin. *Environmental Health Criteria 87.* Geneva, Switzerland: International Programme on Chemical Safety

WSSA (Weed Science Society of America). 1989. *Herbicide Handbook,* 6th edn. Champaign, IL: WSSA

WSSA (Weed Science Society of America). 1994. *Herbicide Handbook,* 7th edn. Champaign, IL: WSSA

CHAPTER 4

KNOWN AND PROBABLE ADVERSE HEALTH IMPACTS

In the United States, there are approximately 2 million agricultural workers, and out of which, according to the Environmental Protection Agency (EPA) estimate, 10,000–20,000 are diagnosed with pesticide poisonings. Other than agricultural workers, individuals working as groundskeepers, pet groomers, fumigators, and a variety of other occupations are at risk for exposure to pesticides, such as fungicides, herbicides, insecticides, rodenticides, and sanitizers (CDC 2013).

PESTICIDE POISONING – SUICIDES/HOMICIDES/ACCIDENTAL

Approximately 800,000 people die globally due to suicide. Among them, around 30% of suicides are due to pesticide poisoning, mostly among rural agricultural areas in low and middle income countries (WHO 2014). In countries with low and middle income, self-poisoning with pesticides is a major public health problem. It was further reported that people dependent on agriculture for livelihood are increasingly committing suicide using pesticides (Alex, Prasad, Kuruvilla, *et al.* 2007).

In Southern India, the suicide rates have been reported to be high (Joseph, Abraham, Muliyil, 2003; Aaron, Joseph, Abraham, *et al.* 2004; Abraham, Abraham, and Jacob 2005; Prasad, Abraham, Minz, *et al.* 2006). From 1986 to 2005, 1741 people committed suicides in a population of about 100,000. Two most common methods used to commit suicide have been hanging (804 of 1741, 46.2%) and poisoning (746 of 1741, 42.8%). Hanging (P < 0.001) was chosen by older people and poisoning by people aged 40 years and younger. More males (465 of 984, 47.3%) preferred poisoning to commit suicide compared to females (281 of 757, 37.1%) (P < 0.001) (Alex, Prasad, Kuruvilla, *et al.* 2007).

Pesticides are common agents used for homicides and suicides in developing countries. Prasad, Abraham, Minz, *et al.* (2007) noted that suicide in low- and middle-income countries is not only a medical and public health problem but is also related to economics and culture. A coordinated and comprehensive response is needed to make any impact (Alex, Prasad, Kuruvilla, *et al.* 2007).

CANCER

The risk of cancer, such as leukaemia and lymphoma, increases in individuals who are exposed to high levels of pesticides. This is especially true for individuals engaged in farming or employed in industry. According to the International Agency for Research into Cancer (IARC), individuals spraying pesticides regularly are at a slightly increased risk of contracting cancer. But for individual pesticides, the evidence was either too weak to come to a conclusion, or only strong enough to suggest a "possible" effect (Cancer Research, UK 2013).

The wider use of pesticides in the community has led to increased controversy. Research has shown that a strong relationship exists between pesticide exposure and cancer development, especially in children. This was found to be particularly true for brain cancer and prostate cancer. Studies have also found that children of workers exposed to pesticides at work are more likely to develop kidney cancer (Bassil, Vakil, Sanborn, *et al.* 2007). This is particularly true for high and prolonged exposures. Specific weaknesses and inherent limitations in epidemiologic studies were noted, particularly around ascertaining whether and how much exposure had taken place. As these studies involved multiple pesticide use, it is felt that exposure to all pesticides should be reduced. The quality of studies examining the association between pesticide use and cancer is variable, consisting mainly of cohort and case control (Bassil, Vakil, Sanborn, *et al.* 2007).

CONGENITAL MALFORMATIONS

García, Benavides, Fletcher, *et al.* (1998) conducted a study in which they found that paternal exposure to pesticides, such as pyridils, aliphatic hydrocarbons, inorganic compounds, and glufosinate, led to a possible risk of congenital malformations. However, it did not find an increased risk for paternal exposure to pesticides in the classes of organophosphates, carbamates, organochlorines, chloroalkylthio fungicides, and organosulphurs. García Benavides, Fletcher, *et al.* (1998) suggested that these findings call for further investigation. In a previous study too, a link between congenital malformations and pesticide exposure was found (Leite, Acosta, and Macchi 2008).

Researchers conducted a study in farm workers to examine the relationship between pesticide exposure and birth defects (Pattison and Toxic Free NC 2006). Different perspectives and methods of assessing the association were adopted in the nine studies following the reviews. The results of these nine studies were not consistent, however, all (except one) indicated the potential association between one or both parents' exposure to pesticides and one or more birth defects (Pattison and Toxic Free NC 2006). The study also emphasizes the need for caution and parental education.

Following are the common birth defects that emerged from the review (Pattison and Toxic Free NC 2006):

- **Cryptorchidism:** A condition in which one or both testes fail to descend normally.
- **Hypospadias:** An abnormality of the penis in which the urethra opens on the underside.
- **Spina bifida:** Defects in the spinal cord and in the vertebrae caused by the incomplete closure of the neural tube.

NEUROLOGICAL DYSFUNCTIONS

Some of the significant concluding remarks of the review paper on the association of pesticide exposure with neurologic dysfunction and disease by Kamel and Hoppin (2004) is given below. There is increasing evidence that chronic exposure to moderate levels of pesticides is neurotoxic and it increases the risk of Parkinson's disease. In order to validate these results, studies in future must consider a detailed assessment of exposure in individuals and the role of genetic susceptibility (Kamel and Hoppin 2004). So far the neurotoxicity studies have found an increase in symptom prevalence and changes in neurobehavioural performance reflecting cognitive and psychomotor dysfunction. However, many of them have found little effect of pesticide exposure on sensory or motor function or direct measures of nerve function (Kamel and Hoppin 2004). Possibility of bias in the increase of symptom prevalence cannot be ruled out. The role of confounders such as injury or neurologic disease in pesticide exposure studies is to be validated to confirm whether the symptoms are due to pesticide exposure or any other clinical reasons. The future studies should attempt to resolve the issue of inconsistencies in pesticide exposure studies (Kamel and Hoppin 2004).

Affirmative results from studies conducted in recent past with more inclusive exposure assessment and the evidence from animal models further support the hypothesis of an association (Kamel and Hoppin 2004). Results for amyotrophic lateral sclerosis and Alzheimer's disease are suggestive but not strong enough to give firm conclusions (Kamel and Hoppin 2004).

NEURODEGENERATIVE DISEASES

Parkinson's Disease

The risk of Parkinson's disease might increase because of exposure to pesticides (Le Couteur, McLean, Taylor, *et al.* 1999). Earlier it was reported that a relation exists between risk of Parkinson's disease and living in rural areas, drinking water from well, and agricultural occupation (Priyadarshi, Khuder, Schaub, *et al.* 2001). More specifically, case-control studies have found that pesticide exposure is associated with increased Parkinson's disease risk, although results are not consistent. Le Couteur, McLean, Taylor, *et al.* (1999) reviewed the studies conducted prior to 1999 and found that 12 out of 20 studies noted a positive association, with 1.6- to 7-fold increase in risk. A few of these studies examined the hazards associated with ever exposure to any pesticide. Based on different study designs on pesticide exposure to neurodegenerative diseases, the following trends were observed.

Case-control studies found the following results:
- Increased risk associated with pesticide use (Baldereschi, Di, Vanni, *et al.* 2003).
- Increased risk associated with cumulative pesticide exposure in occupational settings (Baldi, Cantagrel, Lebailly, *et al.* 2003a; Fall, Fredrikson, Axelson, *et al.* 1999; Herishanu, Medvedovski, Goldsmith, *et al.* 2001).
- Three case-control studies found no association of pesticide exposure with Parkinson's disease (Behari, Srivastava, Das, *et al.* 2001; Kuopio, Marttila, Helenius, *et al.* 1999; Taylor, Saint-Hilaire, Cupples, *et al.* 1999).

Case reports have evidenced the following results:
- Association of Parkinson's disease in individuals exposed to organophosphates (Bhatt, Elias, and Mankodi 1999; Davis, Yesavage, and Berger 1978)
- Parkinson's disease and its relation to exposure of herbicides (as glyphosate) (Barbosa, Leiros da Costa, Bacheschi, *et al.* 2001)

- Paraquat and diquat exposure and prevalence of Parkinson's disease (Sanchez-Ramon, Hefti, and Weiner 1987; Sechi, Agnetti, Piredda, *et al.* 1992)
- Fungicides (asmaneb) and dithiocarbamates exposure and its association with Parkinson's disease (Meco, Bonifati, Vanacore, *et al.* 1994; Hoogenraad 1988)

Cross-sectional study and ecological study evidenced the following trend:

- Association of Parkinsonism with exposure to any pesticide in a cross-sectional study (Engel, Checkoway, Keifer, *et al.* 2001)
- An ecological study found that Parkinson's disease mortality was more in pesticide exposed locations compared to non-pesticide exposed locations (Ritz and Yu 2000)

Alzheimer's Disease

A case-control study associated occupational exposure to unspecified pesticides and fertilizers with the risk of Alzheimer's disease (McDowell, Hill, Lindsay, *et al.* 1994). In a separate smaller study of environmental exposure in the general population, no correlation was noted between herbicides, insecticides, or pesticides and the risk of Alzheimer's disease (Gauthier, Fortier, Courchesne, *et al.* 2001). Occupational exposure to any pesticide evaluated with a job-exposure matrix was associated with twofold increase in the risk of Alzheimer's disease in a cohort of older individuals living in a vineyard-growing region of France and exposed primarily to dithiocarbamate fungicides (Baldi, Lebailly, Mohammed-Brahim, *et al.* 2003b), mild cognitive dysfunction in a population-based prospective study (Bosma, van Boxtel, Ponds, *et al.* 2000), vascular dementia (Lindsay, Hebert, and Rockwood 1997), and risk of dementia among Parkinson's disease patients (Hubble, Kurth, Glatt, *et al.* 1998).

Comprehending the relationship between pesticide exposure and Alzheimer's disease is difficult because in Alzheimer's disease the basic neurochemical defect is the loss of cholinergic neurons. To increase cholinergic tone, Alzheimer's disease is sometimes treated with organophosphate cholinesterase inhibitors (Ringman and Cummings 1999).

ASTHMA

A correlation between exposure to pesticides, bronchial hyper-reactivity, and asthma symptoms have been noted by many clinical and epidemiological studies (Hernandez, Parron, and Alarcon 2011). There are several pesticides that act as sensitizers or irritants and directly damage the bronchial mucosa. This makes the airway very sensitive to allergens or other stimuli. However, many pesticides are only weakly immunogenic. They have a limited potential to sensitize airways in exposed populations. Pesticides are also likely to increase the risk of developing asthma, worsen a previous asthmatic condition, or even cause asthma attacks by increasing bronchial hyper-responsiveness (Hernandez, Parron, and Alarcon 2011).

ENDOCRINE DISRUPTORS

Endocrine disrupting chemicals (EDCs) including pesticides, are compounds which interfere with the normal functioning of the endocrine system of both humans and wildlife (Mnif, Hassine, Bouaziz, *et al.* 2011). A variety of chemicals have been identified as EDCs and several of them are pesticides. Out of the 105 chemicals identified as EDCs, insecticides comprise 46%, herbicides 21%, and fungicides 31%. For a long time some of these EDCs have been withdrawn from general use. However, they are still persistent in the environment in many countries. Examples include dichlorodiphenyltrichloroethane (DDT) and atrazine in many countries.

Endocrine-disrupting chemicals have a strong tendency to bind and activate different hormone receptors, such as oestrogen or androgen receptors. EDCs bind to AR, ER, AhR, PXR, CAR and interfere with the functioning of natural hormones and mimic their action (agonist action) (Tabb and Blumberg 2006). This antagonist action blocks the receptors and inhibits their action. Moreover, EDCs have the potential to interfere with the synthesis, transport, metabolism, and elimination of hormones, which reduces the concentration of natural hormones. Endocrine disruptor pesticides, such as cyhalothrin, amitrole, ioxynil, maneb, mancozeb, fipronil, prodiamine, pentachloronitrobenzene, pyrimethanil, thiazopyr, ziram, and zineb have the potential to inhibit the production of thyroid hormones (Cocco 2002; Akhtar, Kayani, Ahmad, *et al.* 1996; Leghait, Gayrard, Picard-Hagen, *et al.* 2009; Sugiyama, Shimada, Miyoshi, *et al.* 2005). Endocrine disruptor pesticides also interfere with the reproductive and sexual development in humans, and these effects are dependent on many factors, including gender, age, diet, and occupation.

One of the most sensitive factors is age. Compared to adults, human foetuses, infants, and children are more prone to be affected (Birnbaum and Fenton 2003; Goldman, Falk, Landrigan, *et al.* 2004; Sharpe 2006). The harmful effects due to EDCs occur during gametogenesis and the early development of foetus (Sharpe 2006; Sultan, Balaguer, Terouanne, *et al.* 2001; Skakkebaek 2002; Hardell, Bavel, Lindström, *et al.* 2006). But these health effects may not manifest until adulthood (Mnif, Hassine, Bouaziz, *et al.* 2011). Besides, foetuses and infants receive greater doses due to the mobilization of maternal fat reserves during pregnancy and breastfeeding (Anderson and Wolff 2000; Przyrembel, Heinrich-Hirsch, and Vieth 2000). Infants are highly susceptible to pre- and post-natal exposure to endocrine disruptor pesticides. They may suffer from adverse health effects, such as possible lasting impacts on intellectual function (Jacobson and Jacobson 1996; Eskenazi, Marks, Bradman, *et al.* 2006) and delayed adverse effects on the functioning of central nervous system (Ribas-Fito, Cardo, Sala, *et al.* 2003; Beard and Australian Rural Health Research Collaboration 2006).

Similarly, living near agricultural pesticide applications is often cited as a reason for developmental abnormalities in epidemiological studies of low birth weight (Xiang, Nuckols, and Stallones 2000), foetal death (Bell, Hertz-Picciotto, and Beaumont 2001), and childhood cancers (Reynolds, Von Behren, Gunier, *et al.* 2002). Moreover, the occurrence of cryptorchidism and hypospadias (Kristensen, Irgens, Andersen, *et al.* 1997; Carbone, Giordano, Nori, *et al.* 2006) was reported to be higher in places where farming was done extensively and in the male child of women working as gardeners (Weidner, Moller, Jensen, *et al.* 1998). In the recent past, a link has been reported between cryptorchidism and persistent pesticide concentration in maternal breast milk (Skakkebaek 2002; Skakkebaek, Rajpert-De Meyts, and Main 2001; Damgaard, Skakkebaek, Toppari, *et al.* 2006). Adverse effects of pesticide on fertility attributable to their endocrine disrupting activity has also been reported. (Roeleveld and Bretveld 2008).

Exposure to pesticide can lead to other adverse health effects, such as diabetes and cardiovascular toxicity related health problems. In this chapter, different adverse health events are discussed on the basis of systematic reviews and meta-analysis; however, no concrete correlations were observed on their relationship with pesticide exposure. Isolated studies have been avoided because the results are specific to the populations studied and cannot be generalized.

DIABETES

Several studies have found a relation between diabetes and persistent organic pollutants (POPs), including polychlorinated biphenyls (PCBs), polychlorinated dibenzo-p-dioxins, and p,p'-diphenyldichloroethene (DDE). These POPs have been reported in most American population at relatively low levels (CDC 2005). In many cross-sectional researches, a link has been found between diabetes and dioxin-like chemicals, non-dioxin-like PCBs, DDE, and other organochlorine pesticides (Codru, Schymura, Negoita, *et al.* 2007; Cox, Niskar, Narayan, *et al.* 2007; Everett, Frithsen, Diaz, *et al.* 2007; Fierens, Mairesse, Heilier, *et al.* 2003; Lee, Lee, Steffes, *et al.* 2007b; Persky, McCann, Mallin, *et al.* 2002; Radikova, Koska, Ksinantova, *et al.* 2004; Rylander, Rignell-Hydbom, and Hagmar 2005; Uemura, Arisawa, Hiyoshi, *et al.* 2008).

In a study by Morgan, Lin, and Saikaly (1980) during 1971–73, the serum DDE and *p,p'*-DDT levels in workers exposed to pesticides were measured and 3669 workers were followed during 1977–78 for newly diagnosed diabetes. Serum DDE and DDT levels were found to be extremely high among workers with incidence of diabetes in comparison to workers who did not have diabetes. For 25 years, Vasiliu, Cameron, Gardiner, *et al.* (2006) studied 1384 adults prospectively. They noted that PCBs were associated with incidence of diabetes in women but not in men and remained associated when cases occurring during the first 15 years of follow-up were excluded, suggesting that reverse causality was an unlikely explanation for the relationship. Investigations of veterans of Operation Ranch Hand, who were exposed to 2,3,7,8-tetrachlorodibenzo-*p*-dioxin (TCDD) through the application of herbicides in Vietnam, found association of TCDD exposure with incidence of diabetes (Henriksen, Ketchum, Michalek, *et al.* 1997).

The evidence is more consistent for lower level exposures (Everett, Frithsen, Diaz, *et al.* 2007; Fierens, Mairesse, Heilier, *et al.* 2003; Henriksen, Ketchum, Michalek, *et al.* 1997; Lee, Lee, Steffes, *et al.* 2007b; Longnecker and Michalek 2000; Uemura, Arisawa, Hiyoshi, *et al.* 2008) than for higher exposures, particularly for high-level TCDD exposures in workers engaged in the manufacture of TCDD contaminated herbicides (Longnecker and Daniels 2001). POPs is affected by changes in body fat (Wolff, Britton, Teitelbaum, *et al.* 2005).

Scientists have studied different mechanisms by which POPs, especially dioxin-like chemicals, risk the incidence of diabetes. These include hyperinsulinemia, antagonism of peroxisome proliferator-activated receptor-γ (PPAR-γ) expression, induction of tumour necrosis factor-α (TNF-α), autoimmunity, and alterations in steroid metabolism (Turyk, Anderson, Knobeloch, *et al.* 2009). In individuals without diabetes, it was observed that dioxin exposure was related to hyperinsulinemia (Cranmer,

Louie, Kennedy, *et al.* 2000; Henriksen, Ketchum, Michalek, *et al.* 1997; Lee, Lee, Steffes, *et al.* 2007a). Dioxin caused the following effects:

- Suppressed PPAR-γ protein expression (Cimafranca, Hanlon, and Jefcoate 2004)
- Induced TNF-α secretion
- Decreased glucose transport and lipoprotein lipase activity (Kern, Dicker-Brown, Said, *et al.* 2002)
- Decreased glucose transporter activity in mice (Liu and Matsumura 1995)

Workers exposed to PCB showed elevated levels of autoantibodies that are linked with diabetes (Langer, Tajtakova, Guretzki, *et al.* 2002). Although DDE is reported to have anti-androgenic effects, the mechanisms responsible for DDE effects on diabetes have not been well researched. On the other hand, DDT and other metabolites of DDE are estrogenic (ASTDR 2002). New research suggests that testosterone may differentially modulate risk of diabetes in men and women. In contrast, sex hormone-binding globulin decreases risk in both men and women (Ding, Song, Malik, *et al.* 2006).

REPRODUCTIVE ABNORMALITIES

The data collected for the study of reproductive toxicity in couples engaged in spraying organochlorine, organophosphorus, and carbamate insecticides in cotton fields were analysed, and they revealed abnormal reproductive performance in the couples (Rupa, Reddy, and Reddy 1991). Male children exposed to endosulfan showed delayed sexual maturity and interferences with sex hormone synthesis (Saiyed, Dewan, Bhatnagar, *et al.* 2003). Table 1 gives a comparison of reproductive effects observed in couples engaged in spraying organochlorine, organophosphorus, and carbamate insecticides (exposed column) in cotton fields with unexposed subjects.

Table 1 Comparison of reproductive effects observed in couples engaged in spraying insecticides with unexposed subjects

Reproductive problems	Exposed (%)	Unexposed (%)
Abortions	26.0	15.0
Stillbirths	8.7	2.6
Neonatal deaths	9.2	2.2
Congenital defects	3.0	0.1

Source Rupa, Reddy, and Reddy (1991)

DIRECT EFFECTS ON PESTICIDE HANDLERS

Pesticide retail shops are predominantly located in urban areas and they cater to the demand of urban population for household products used in lawns and gardens. The reports dealing with the health concerns of pesticide exposure are confined to either incidental/accidental or occupational poisoning (Maria, Dario, and Eloisa 2006; Patricia, Partanen, Wesseling, *et al.* 2005; Harley, Marks, Bradman, *et al.* 2008; Kesavachandran, Singh, Mathur, *et al.* 2006a; Kesavachandran, Rastogi, and Mathur 2006b; Srivastava, Gupta, Bihari, *et al.* 1995; Srivastava, Gupta, Bihari, *et al.* 2000). Individuals who work in the pesticide manufacturing industry and pesticide shops and those who use pesticides for agricultural and other public health activities are exposed to pesticides. In India, pesticide retailers play an important part in the pesticide distribution and hardly using personal protective equipment (PPE) while handling pesticides. Moreover, as India's population is huge, this group comprises a considerable number of subjects.

No recorded data are available on health problems among shopkeepers occupationally exposed to pesticides in India and other developing countries. In India and other developing countries, it is through retail shops that pesticides are distributed for agricultural purposes, and hence such shopkeepers are exposed to a mixture of pesticides. There is a need to protect such occupations, undertake detailed studies for eliciting specific issues, and formulate and implement regulatory guidelines in pesticide retail marketing. For more details on pesticide-retail shops in north India, refer to Kesavachandran, Pathak, Fareed, *et al.* (2009). Shops sell different kinds of pesticides, including organophosphates, organochlorines, carbamates, pyrethroids, and fumigants. Usually pesticide shops have a showroom and a warehouse. While in the showroom limited quantities of pesticides are displayed, in a warehouse large quantities of the same are dumped on the floor. In the showroom, plastic packets, bottles, and tin cans containing pesticides are kept in different racks. With no exhaust facility and only one door opening towards the showroom, warehouses have poor ventilation. In some units, both the showroom and warehouse are in the same shop. In the study conducted, 10% of the subjects were owners who would mostly confine to showrooms. The remaining workers were those who shuttled between warehouse and showroom in course of their work (Kesavachandran, Pathak, Fareed, *et al.* 2009).

"Individuals handling pesticides get exposed via ingestion, inhalation, and dermal contact. As these individuals do not use PPE, the exposure occurs during spillage, loading, unloading, while keeping pesticides in racks, among others. Non-use of PPE while handling pesticides is primarily attributed to poor enforcement of regulations and lack of proper legislative policies with respect to safe and proper handling. The level of pesticide exposure is largely dependent on PPE use, hygiene practice, spills, and attitudes towards risk. All these factors should be considered in the estimation of exposure. However, the relation of these determinants to exposure is complex" (Kamel and Hoppin 2004).

Shopkeepers showed various nervous system related symptoms, such as headache, changes in smell/taste, fatigue, memory loss, tremors, frequent fainting, and pain in limbs (Kesavachandran, Pathak, Fareed, *et al.* 2009). Pesticides-exposed subjects showed cognitive, motor, and sensory dysfunctions (Kamel and Hoppin 2004). Symptoms observed in them included headache, dizziness, fatigue, insomnia, nausea, chest tightness, difficulty in breathing, confusion, difficulty in concentrating, weakness, tremor, numbness, tingling, and visual disturbance (Kamel and Hoppin 2004). Different studies gave different results with regard to pesticide effects. While one study showed that the sense of smell was not affected by organophosphates (Steenland, Dick, Howell, *et al.* 2000), another study indicated a relationship between sense of smell and fumigants (Calvert, Mueller, Fajen, *et al.* 1998). A report found an association between exposure to multiple pesticides and tremor (Kamel and Hoppin 2004). However, the same result was not observed when exposure was limited to organophosphate in another study (Steenland, Dick, Howell, *et al.* 2000).

Nerve conduction studies showed significant decline in motor nerve conduction velocity among pesticide sprayers (Pathak, Fareed, Bihari, *et al.* 2011). Nerve conduction velocity is an important aspect of nerve conduction study. It indicates the effect of pesticides on the peripheral nervous system (Kimura, Yokoyama, Sato, *et al.* 2005). In an earlier study, workers exposed to organophosphates (Mishra, Nag, Khan, *et al.* 1988) and fumigators (Calvert, Mueller, Fajen, *et al.* 1998) showed reduced nerve conduction velocity. In another study, complete opposite results were witnessed. Nerve conduction was little impaired in subjects exposed to organophosphates (Kamel and Hoppin 2004; Steenland, Dick, Howell, *et al.* 2000).

"Organophosphorus compounds are able to easily move across the blood–brain barrier. This property allows accumulation of excess amounts of acetylcholine at particular central nervous system receptors and at neuromuscular junctions by inhibiting cholinesterase. Neurotoxic effects resulting from exposure to organophosphorus compounds occur in three successive clinical stages. These are acute cholinergic crisis, intermediate syndrome, and organophosphate-induced delayed peripheral neuropathy" (Zorzon, Capus, Pellegrino, *et al.* 2002).

Some of the chemicals that lead to these distinctive intoxications are fenthion, dimethoate, monocrotophos, and methamidophos (Kimura, Yokoyama, Sato, *et al.* 2005). In comparison to control group, the exposed shopkeepers group showed higher gastrointestinal problems, such as hyperacidity, burning sensation in abdomen, pain in abdomen, vomiting, constipation, and jaundice (Kesavachandran, Pathak, Fareed, *et al.* 2009). Scientists have reported gastrointestinal problems in pesticide sprayers too (Kesavachandran, Rastogi, and Mathur 2006b; Singh, Jyoti, Kesavachandran, *et al.* 2007).

Pesticides also get ingested when users do not wash their hands properly before eating. At the gastrointestinal tract, pesticides can be absorbed at any portions, with small intestine being the major absorption site (ITRC 1990). Shopkeepers exposed to pesticides and 15% of pesticides sprayers also showed musculoskeletal disorders, such as swollen joints, bodyache, muscle twitching, and weakness in arms and legs (Kesavachandran,Pathak, Fareed, *et al.* 2009; ITRC 1990). Subjects may suffer from musculoskeletal problems because of not following proper ergonomic practices when loading and unloading pesticide containers (Kesavachandran,Pathak, Fareed, *et al.* 2009).

Shopkeepers exposed to pesticides showed dermal symptoms such as warm and burning sensation of skin and itching. The subjects can also suffer from ocular problems (Kesavachandran, Rastogi, and Mathur 2006b; ITRC 1990). Ocular toxicity takes place when unprotected eyes are exposed to pesticides. The ocular tissues then absorb these chemicals, which enter the circulation (Jaga and Dharmani 2006). Accidental splashes of chemicals can also result in exposure (Jaga and Dharmani 2006). Ocular problems include watering of eyes, pain, swollen red eye, and irritation.

In comparison to control subjects, the exposed subjects showed significant reduction in the mean values of lung functions, forced vital capacity (FVC) and peak expiratory flow rate (PEFR). Lung function abnormalities were observed among pesticide retailers occupationally exposed to pesticides (Kesavachandran, Pathak, Fareed, *et al.* 2009), pesticide sprayers (Kesavachandran, Singh, Mathur, *et al.* 2006a; Rastogi, Gupta, Husain, *et al.* 1989), and among South African farm workers (Dalvie, White, Raine, *et al.* 1999). Fumigants viz., methyl bromide and sulphur dioxide, can cause respiratory health problems in occupationally exposed agrarian subjects (Popendorf, Donham, Easton, *et al.* 1985). Earlier report suggests that paraquat exposure is associated with respiratory disease such as increased wheezing among Nicaraguan banana workers (Castro-Gutierrez, McConnell, Andersson, *et al.* 1997).

REFERENCES

Aaron, R., A. Joseph, S. Abraham, J. Muliyil, K. George, J. Prasad, S. Minz, V. J. Abraham, and A. Bose. 2004. Suicides in young people in rural southern India. *Lancet* 363(1): 117–118

Abraham, V. J., S. Abraham, and K. S. Jacob. 2005. Suicide in the elderly in Kaniyambadi block, Tamil Nadu, South India. *International Journal of Geriatric Psychiatry* 20: 953–955

Akhtar, N., S. A. Kayani, M. M. Ahmad, and M. Shahab. 1996. Insecticide-induced changes in secretory activity of the thyroid gland in rats. *Journal of Applied Toxicology* 16: 397–400

Alex, R., J. Prasad, A. Kuruvilla, and K. S. Jacob. 2007. Self-poisoning with pesticides in India. *The British Journal of Psychology* 190: 274–275

Anderson, H. and M. S. Wolff. 2000. Environmental contaminants in human milk. *Journal of Exposure Analysis and Environmental Epidemiology* 10: 755–760

ASTDR (Agency for Toxic Substances and Disease Registry). 2002. *Toxicological Profile for DDT, DDE, and DDD*. Atlanta, GA: ASTDR

Baldereschi, M., C. A. Di, P. Vanni, A. Ghetti, P. Carbonin, L. Amaducci, D. Inzitari, Italian Longitudinal Study on Aging Working Group. 2003. Lifestyle-related risk factors for Parkinson's disease: a population-based study. *Acta Neurologica Scandinavica* 108: 239–244

Baldi, I., A. Cantagrel, P. Lebailly, F. Tison, B. Dubroca, V. Chrysostome, J. F. Dartigues, and P. Brochard. 2003a. Association between Parkinson's disease and exposure to pesticides in southwestern France. *Neuroepidemiology* 22: 305–310

Baldi, I., P. Lebailly, B. Mohammed-Brahim, L. Letenneur, J. F. Dartigues, and P. Brochard. 2003b. Neurodegenerative diseases and exposure to pesticides in the elderly. *American Journal of Epidemiology* 157: 409–414

Barbosa, E. R., M. D. Leiros da Costa, L. A. Bacheschi, M. Scaff, and C. C. Leite. 2001. Parkinsonism after glycine-derivate exposure. *Movement Disorders* 16: 565–568

Bassil, K. L., C. Vakil, M. Sanborn, D. C. Cole, J. S. Kaur, and K. J. Kerr. 2007. Cancer health effects of pesticides: systematic review. *Canadian Family Physician* 53: 1704–1711

Beard, J. and Australian Rural Health Research Collaboration. 2006. DDT and human health. *Science of the Total Environment* 355: 78–89

Behari, M., A. K. Srivastava, R. R. Das, and R. M. Pandey. 2001. Risk factors of Parkinson's disease in Indian patients. *Journal of the Neurological Sciences* 190: 49–55

Bell, E. M., I. Hertz-Picciotto, and J. J. Beaumont. 2001. A case-control study of pesticides and fetal death due to congenital anomalies. *Epidemiology* 12: 148–156

Bhatt, M. H., M. A. Elias, and A. K. Mankodi. 1999. Acute and reversible Parkinsonism due to organophosphate pesticide intoxication: five cases. *Neurology* 52: 1467–1471

Birnbaum, L. S. and S. E. Fenton. 2003. Cancer and developmental exposure to endocrine disruptors. *Environmental Health Perspectives* 111: 389–394

Bosma, H., M. P. van Boxtel, R. W. Ponds, P. J. Houx, and J. Jolles. 2000. Pesticide exposure and risk of mild cognitive dysfunction. *Lancet* 356: 912–913

Calvert, G. M., C. A. Mueller, J. M. Fajen, D. W. Chrislip, J. Russo, T. Briggle, L. E. Fleming, A. J. Suruda, and K. Steenland. 1998. Health effects associated with sulfuryl fluoride and methyl bromide exposure among structural fumigation workers. *American Journal of Public Health* 88: 1774–1780

Cancer Research UK. 2013. Pesticides and cancer. Details available at <http://www.cancerresearchuk.org/cancer-info/healthyliving/cancercontroversies/pesticides/pesticides-and-cancer>, last accessed on 4 March 2011

Carbone, P., F. Giordano, F. Nori, A. Mantovani, D. Taruscio, L. Lauria, and I. Figà-Talamanca. 2006. Cryptorchidism and hypospadias in the Sicilian district of Ragusa and the use of pesticides. *Reproductive Toxicology* 22: 8–12

Castro-Gutierrez, N., R. McConnell, K. Andersson, F. Pacheco-Anton, and C. Hogstedt. 1997. Respiratory symptoms, spirometry and chronic occupational paraquat exposure. *Scandinavian Journal of Work, Environment and Health* 23: 421–427

CDC (Center for Disease Control and Prevention). 2005. *Third National Report on Human Exposure to Environmental Chemicals.* Atlanta, GA: CDC

CDC (Center for Disease Control and Prevention). 2013. Pesticide illness and injury surveillance. Details available at <www.cdc.gov/niosh/topics/pesticides>, last accessed on 13 December 2011

Cimafranca, M. A., P. R. Hanlon, and C. R. Jefcoate. 2004. TCDD administration after the pro-adipogenic differentiation stimulus inhibits PPAR gamma through a MEK-dependent process but less effectively suppresses adipogenesis. *Toxicology and Applied Pharmacology* 196: 156–168

Cocco, P. 2002. On the rumors about the silent spring. Review of the scientific evidence linking occupational and environmental pesticide exposure to endocrine disruption health effects. *Cadernos de SaúdePública* 18: 379–402

Codru, N., M. J. Schymura, S. Negoita, Akwesasne Task Force on the Environment, R. Rej, and D. O. Carpenter. 2007. Diabetes in relation to serum levels of polychlorinated biphenyls and chlorinated pesticides in adult Native Americans. *Environmental Health Perspectives* 115: 1442–1447

Cox, S., A. S. Niskar, K. M. Narayan, and M. Marcus. 2007. Prevalence of self-reported diabetes and exposure to organochlorine pesticides among Mexican Americans: Hispanic health and nutrition examination survey, 1982–1984. *Environmental Health Perspectives* 115: 1747–1752

Cranmer, M., S. Louie, R. H. Kennedy, P. A. Kern, and V. A. Fonseca. 2000. Exposure to 2,3,7,8-tetrachlorodibenzo-*p*-dioxin (TCDD) is associated with hyperinsulinemia and insulin resistance. *Toxicological Sciences* 56: 431–436

Dalvie, M. A., N. White, R. Raine, J. E. Myers, L. London, M. Thompson, and D. C. Christiani. 1999. Long-term respiratory health effects of the herbicide, paraquat, among workers in the Western Cape. *Occupational and Environmental Medicine* 56: 391–396

Damgaard, I. N., N. E. Skakkebaek, J. Toppari, H. E. Virtanen, H. Shen, K. W. Schramm, J. H. Petersen, T. K. Jensen, and K. M. Main. 2006. Persistent pesticides in human breast milk and cryptorchidism. *Environmental Health Perspectives* 114: 1133–1138

Davis, K., J. Yesavage, and P. Berger. 1978. Single case study. Possible organophosphate-induced parkinsonism. *The Journal of Nervous and Mental Disease* 166: 222–225

Ding, E. L., Y. Song, V. S. Malik, and S. Liu. 2006. Sex differences of endogenous sex hormones and risk of type 2 diabetes: a systematic review and meta-analysis. *JAMA* 295: 1288–1299

Engel, L. S., H. Checkoway, M. C. Keifer, N. S. Seixas, W. T. Longstreth, Jr, K. C. Scott, K. Hudnell, W. Anger, and R. Camicioli. 2001. Parkinsonism and occupational exposure to pesticides. *Occupational and Environmental Medicine* 58: 582–589

Eskenazi, B., A. R. Marks, A. Bradman, L. Fenster, C. Johnson, and D. B. Barr. 2006. In utero exposure to dichlorodiphenyltrichloroethane (DDT) and dichlorodiphenyldichloroethylene (DDE) and neurodevelopment among young Mexican American children. *Pediatrics* 118: 233–241

Everett, C. J., I. L. Frithsen, V. A. Diaz, R. J. Koopman, W. M. Simpson, Jr, A. G. Mainous, 3rd. 2007. Association of a polychlorinated dibenzo-*p*-dioxin, a polychlorinated biphenyl, and DDT with diabetes in the 1999–2002 National Health and Nutrition Examination Survey. *Environmental Research* 103: 413–418

Fall, P. A., M. Fredrikson, O. Axelson, and A. K. Granerus. 1999. Nutritional and occupational factors influencing the risk of Parkinson's disease: a case-control study in southeastern Sweden. *Movement Disorders* 14: 28–37

Fierens, S., H. Mairesse, J. F. Heilier, C. De Burbure, J. Focant, G. Eppe, E. De Pauw, and A. Bernard. 2003. Dioxin/polychlorinated biphenyl body burden, diabetes and endometriosis: findings in a population-based study in Belgium. *Biomarkers* 8: 529–534

García, A. M., F. G. Benavides, T. Fletcher, and E. Orts. 1998. Paternal exposure to pesticides and congenital malformations. *Scandinavian Journal of Work, Environment and Health* 24(6): 473–480

Gauthier, E., I. Fortier, F. Courchesne, P. Pepin, J. Mortimer, and D. Gauvreau. 2001. Environmental pesticide exposure as a risk factor for Alzheimer's disease: a case-control study. *Environmental Research* 86: 37–45

Goldman, L., H. Falk, P. J. Landrigan, S. J. Balk, R. Reigart, and R. A. Etzel. 2004. Environmental pediatrics and its impact on government health policy. *Pediatrics* 113: 1146–1157

Hardell, L., B. Bavel, G. Lindström, M. Eriksson, and M. Carlberg. 2006. In utero exposure to persistent organic pollutants in relation to testicular cancer risk. *International Journal of Andrology* 29: 228–234

Harley, K. G., A. R. Marks, A. Bradman, D. B. Barr, and B. Eskenazi. 2008. DDT exposure, work in agriculture and time to pregnancy among farm workers in California. *Journal of Occupational and Environmental Medicine* 50: 1335–1342

Henriksen, G. L., N. S. Ketchum, J. E. Michalek, and J. A. Swaby. 1997. Serum dioxin and diabetes mellitus in veterans of Operation Ranch Hand. *Epidemiology* 8: 252–258

Herishanu, Y. O., M. Medvedovski, J. R. Goldsmith, and E. Kordysh. 2001. A case-control study of Parkinson's disease in urban population of southern Israel. *Canadian Journal of Neurological Sciences* 28: 144–147

Hernandez, A. F., T. Parron, and R. Alarcon. 2011. Pesticides and asthma. *Current Opinion in Allergy and Clinical Immunology* 11(2): 90–96

Hoogenraad, T. 1988. Dithiocarbamates and Parkinson's disease [Letter]. *Lancet* 1: 767

Hubble, J. P., J. H. Kurth, S. L. Glatt, M. C. Kurth, G. D. Schellenberg, R. E. Hassanein, A. Lieberman, and W. C. Koller. 1998. Gene-toxin interaction as a putative risk factor for Parkinson's disease with dementia. *Neuroepidemiology* 17: 96–104

ITRC (Industrial Toxicology Research Centre). 1990. Epidemiological health survey of pesticide sprayers (a report). Industrial Toxicology Research Centre, New Delhi

Jacobson, J. L. and S. W. Jacobson. 1996. Intellectual impairment in children exposed to polychlorinated biphenyls in utero. *The New England Journal of Medicine* 335: 783–789

Jaga, K. and C. Dharmani. 2006. Ocular toxicity from pesticide exposure: a recent review. *Environmental Health and Preventive Medicine* 11: 102–107

Joseph, A., S. Abraham, J. P. Muliyil, K. George, J. Prasad, S. Minz, V. J. Abraham, and K. S. Jacob. 2003. Evaluation of suicide rates in rural India using verbal autopsies, 1994–99. *BMJ* 326: 1121–1122

Kamel, F. and J. A. Hoppin. 2004. Association of pesticide exposure with neurologic dysfunction and disease. *Environmental Health and Perspectives* 112: 950–958

Kern, P. A., A. Dicker-Brown, S. T. Said, R. Kennedy, and V. A. Fonseca. 2002. The stimulation of tumor necrosis factor and inhibition of glucose transport and lipoprotein lipase in adipose cells by 2,3,7,8-tetrachlorodibenzo-*p*-dioxin. *Metabolism* 51: 65–68

Kesavachandran, C., M. K. Pathak, M. Fareed, V. Bihari, N. Mathur, and A. K. Srivastava. 2009. Health risks of employees working in pesticide retail shops: An exploratory study. *Indian Journal of Occup. Env. Med.* 13(3): 121–126

Kesavachandran, C., S. K. Rastogi, and N. Mathur. 2006b. Health status among pesticide applicators at a mango plantation in India. *Journal of Pest Science* 8: 1–9

Kesavachandran, C., V. K. Singh, N. Mathur, S. K. Rastogi, M. K. Siddiqui, M. M. Reddy, R. S. Bharti, and A. M. Khan. 2006a. Possible mechanism of pesticide toxicity related oxidative stress leading to airway narrowing. *Redox Report* 11: 159–162

Kimura, K., K. Yokoyama, H. Sato, R. B. Nordin, L. Naing, S. Kimura, S. Okabe, T. Maeno, Y. Kobayashi, F. Kitamura, and S. Araki. 2005. Effects of pesticides on the peripheral and central nervous system in tobacco farmers in Malaysia: studies on peripheral nerve conduction, brain-evoked potentials, and computerized posturography. *Industrial Health* 43: 285–294

Kristensen, P., L. M. Irgens, A. Andersen, A. S. Bye, and L. Sundheim. 1997. Birth defects among offspring of Norwegian farmers, 1967–1991. *Epidemiology* 8: 537–544

Kuopio, A. M., R. J. Marttila, H. Helenius, and U. K. Rinne. 1999. Environmental risk factors in Parkinson's disease. *Movement Disorders* 14: 928–939

Langer, P., M. Tajtakova, H. Guretzki, A. Kocan, J. Petrik, J. Chovancova, B. Drobná, S. Jursa, M. Pavúk, T. Trnovec, E. Seböková, and I. Klimes. 2002. High prevalence of anti-glutamic acid decarboxylase (anti-GAD) antibodies in employees at a polychlorinated biphenyl production factory. *Archives of Environmental Health* 57: 412–415

Le Couteur, D. G., A. J. McLean, M. C. Taylor, B. L. Woodham, and P. G. Board. 1999. Pesticides and Parkinson's disease. *Biomedicine and Pharmacotherapy* 53: 122–130

Lee, D. H., I. K. Lee, M. Steffes, and D. R. Jacobs, Jr. 2007a. Association between serum concentrations of persistent organic pollutants and insulin resistance among nondiabetic adults. *Diabetes Care* 30: 622–628

Lee, D. H., I. K. Lee, M. Steffes, and D. R. Jacobs, Jr. 2007b. Extended analyses of the association between serum concentrations of persistent organic pollutants and diabetes. *Diabetes Care* 30: 1596–1598

Leghait, J., V. Gayrard, N. Picard-Hagen, M. Camp, E. Perdu, P. L. Toutain, and C. Viguié. 2009. Fipronil-induced disruption of thyroid function in rats is mediated by increased total and free thyroxine clearances concomitantly to increased activity of hepatic enzymes. *Toxicology* 255: 38–44

Leite, S. B., M. Acosta, and M. L. Macchi. 2008. Congenital malformations associated with pesticides in Encarnación, Paraguay. *Pediatrics* 121: S107

Lindsay, J., R. Hebert, and K. Rockwood. 1997. The Canadian study of health and aging: risk factors for vascular dementia. *Stroke* 28: 526–530

Liu, P. C. and F. Matsumura. 1995. Differential effects of 2,3,7,8-tetrachlorodibenzo-*p*-dioxin on the "adipose-type" and "brain-type" glucose transporters in mice. *Molecular Pharmacology* 47: 65–73

Longnecker, M. P. and J. E. Michalek. 2000. Serum dioxin level in relation to diabetes mellitus among Air Force veterans with background levels of exposure. *Epidemiology* 11: 44–48

Longnecker, M. P. and J. L. Daniels. 2001. Environmental contaminants as etiologic factors for diabetes. *Environmental Health Perspectives* 109: 871–876

Maria, C. P., X. P. Dario, and D. C. Eloisa. 2006. Acute poisoning with pesticides in the state of Mato, Grosso do Sul, Brazil. *Science of the Total Environment* 357: 88–95

McDowell, I., G. Hill, J. Lindsay, B. Helliwell, L. Costa, and L. Beattie.1994. The Canadian study of health and aging: risk factors for Alzheimer's disease in Canada. *Neurology* 44:2073–2080

Meco, G., V. Bonifati, N. Vanacore, and E. Fabrizio. 1994. Parkinsonism after chronic exposure to the fungicide maneb (manganese ethylene-bis-dithiocarbamate). *Scandinavian Journal of Work, Environment and Health* 20: 301–305

Mishra, U. K., D. Nag, W. A. Khan, and P. K. Ray. 1988. A study of nerve conduction velocity, late responses and neruo muscular synapse functions in organophosphate workers in India. *Archives of Toxicology* 61: 496–500

Mnif, W., A. I. H. Hassine, A. Bouaziz, A. Bartegi, O. Thomas, and B. Roig. 2011. Effect of endocrine disruptor pesticides: a review. *International Journal of Environmental Research Public Health* 8: 2265–2303

Morgan, D. P., L. I. Lin, and H. H. Saikaly. 1980. Morbidity and mortality in workers occupationally exposed to pesticides. *Archives of Environmental Contamination and Toxicology* 9: 349–382

Pathak, M. K., M. Fareed, V. Bihari, N. Mathur, D. K. Patel, M. M. K. Reddy, and C. Kesavachandran. 2001. Nerve conduction studies in sprayers occupationally exposed to mixture of pesticides in a mango plantation at Lucknow, North India. Toxicol Environ. Chem 93(1): 188–196

Patricia, M., T. Partanen, C. Wesseling, V. Bravo, C. Ruepert, and I. Burstyn. 2005. Assessment of pesticide exposure in the agricultural population of Costa Rica. *The Annals of Occupational Hygiene* 49: 375–384

Pattison, F. and Toxic Free NC. 2006. Examining the evidence on pesticide exposure and birth defects in farmworkers: an annotated bibliography, with resources for lay readers. Details available at <www.toxicfreenc.org/informed/pdfs/Evidence_May06_08.pdf>last accessed on 24 July 2011

Persky, V., K. McCann, K. Mallin, S. Freels, J. Piorkowski, and L. K. Chary. 2002. The La Salle Electrical Utilities Company Morbidity Study I. ATSDR Monograph PB02-100121. Atlanta, GA: Agency for Toxic Substances and Disease Registry

Popendorf, W., K. J. Donham, D. N. Easton, and J. Silk. 1985. A synopsis of agricultural respiratory hazards. *American Industrial Hygiene Association Journal* 46: 154–161

Prasad, J., V. J. Abraham, S. Minz, S. Abraham, A. Joseph, J. P. Muliyil, K. George, and K. S. Jacob. 2006. Rates and factors associated with suicide in Kaniyambadi Block, Tamil Nadu, South India, 2000–02. *International Journal of Social Psychiatry* 52: 65–71

Priyadarshi, A., S. A. Khuder, E. A. Schaub, and S. S. Priyadarshi. 2001. Environmental risk factors and Parkinson's disease: a metaanalysis. *Environmental Research* 86: 122–127

Przyrembel, H., B. Heinrich-Hirsch, and B. Vieth. 2000. Exposition to and health effects of residues in human milk. *Advances in Experimental Medicine and Biology* 478: 307–325

Radikova, Z., J. Koska, L. Ksinantova, R. Imrich, A. Kocan, J. Petrik, M. Huckova, L. Wsolova, P. Langer, T. Trnovec, E. Sebokova, and I. Klimes. 2004. Increased frequency of diabetes and other forms of dysglycemia in the population of specific areas of eastern Slovakia chronically exposed to contamination with polychlorinated biphenyls (PCB). *Organohalogen Compounds* 66: 3547–3551

Rastogi, S. K., B. N. Gupta, T. Husain, N. Mathur, and N. Garg. 1989. Study on respiratory impairment among pesticide sprayers in mango plantations. *American Journal of Industrial Medicine* 16: 529–538

Reynolds, P., J. Von Behren, R. B. Gunier, D. E. Goldberg, A. Hertz, and M. E. Harnly. 2002. Childhood cancer and agricultural pesticide use: an ecologic study in California. *Environmental Health Perspectives* 110: 319–324

Ribas-Fito, N., E. Cardo, M. Sala, M. Eulalia de Muga, C. Mazon, A. Verdu, M. Kogevinas, J. O. Grimalt, and J. Sunyer. 2003. Breastfeeding exposure to organochlorine compounds and neurodevelopment in infants. *Pediatrics* 111: 580–585

Ringman, J. M. and J. L. Cummings. 1999. Metrifonate: update on a new antidementia agent. *Journal of Clinical Psychiatry* 60: 776–782

Ritz, B. and F. Yu. 2000. Parkinson's disease mortality and pesticide exposure in California 1984–1994. *International Journal of Epidemiology* 292: 323–329

Roeleveld, N. and R. Bretveld. 2008. The impact of pesticides on male fertility. *Current Opinion in Obstetrics and Gynecology* 20: 229–233

Rupa, D. S., P. P. Reddy, and O. S. Reddy. 1991. Reproductive performance in population exposed to pesticides in cotton fields in India. *Environmental Research* 55: 23–126

Rylander, L., A. Rignell-Hydbom, and L. Hagmar. 2005. A cross-sectional study of the association between persistent organochlorine pollutants and diabetes. *Environmental Health* 4: 28

Saiyed, H. N., A. Dewan, V. Bhatnagar, V. Shenoy, R. Shenoy, H. Rajmohan, K. Patel, R. Kashyap, P. Kulkarni, B. Rajan, and B. Lakkad. 2003. Effect of endosulfan on male reproductive development. *Environmental Health Perspectives* 111(16): 1958–1962

Sanchez-Ramon, J., F. Hefti, and W. I. Weiner. 1987. Paraquat and Parkinson disease [Letter]. *Neurology* 37: 728

Sechi, G., V. Agnetti, M. Piredda, M. Canu, F. Deserra, H. A. Omar, and G. Rosati. 1992. Acute and persistent parkinsonism after use of diquat. *Neurology* 42: 261–263

Sharpe, R. R. M. 2006. Pathways of endocrine disruption during male sexual differentiation and masculinization. *Best Practice and Research Clinical Endocrinology and Metabolism* 20: 91–110

Singh, V. K., R. M. M. Jyoti, C. Kesavachandran, S. K. Rastogi, and M. K. Siddiqui. 2007. Biomonitoring of organochlorines, glutathione, lipid peroxidation and cholinesterase activity among pesticide sprayers in mango orchards. *Clinica Chimica Acta* 377: 268–272

Skakkebaek, N. N. 2002. Endocrine disrupters and testicular dysgenesis syndrome. *Hormone Research* 57: 43

Skakkebaek, N. E., E. Rajpert-De Meyts, and K. M. Main. 2001. Testicular dysgenesis syndrome: an increasingly common developmental disorder with environmental aspects. *Human Reproduction* 16: 972–978

Srivastava, A. K., B. N. Gupta, V. Bihari, N. Mathur, B. S. Pangtey, R. S. Bharti, and M. M. Godbole. 1995. Organochlorine pesticide exposure and thyroid function: a study in human subjects. *Journal of Environmental Pathology, Toxicology and Oncology* 14: 107–110

Srivastava, A. K., B. N. Gupta, V. Bihari, N. Mathur, L. P. Srivastava, B. S. Pangtey, and P. Kumar. 2000. Clinical, biochemical and neurobehavioural studies of workers engaged in the manufacture of quinalphos. *Food and Chemical Toxicology* 38: 65–69

Steenland, K., R. B. Dick, R. J. Howell, D. W. Chrislip, C. J. Hines, T. M. Reid, E. Lehman, P. Laber, E. F. Krieg, Jr, and C. Knott. 2000. Neurologic function among termiticide applicators exposed to chlorpyrifos. *Environmental Health Perspectives* 108: 293–300

Sugiyama, S., N. Shimada, H. Miyoshi, and K. Yamauchi. 2005. Detection of thyroid system disrupting chemicals using in vitro and in vivo screening assays in *Xenopus laevis. Toxicological Sciences* 88: 367–374

Sultan, C., P. Balaguer, B. Terouanne, V. Georget, F. Paris, C. Jeandel, S. Lumbroso, and J. Nicolas. 2001. Environmental xenoestrogens, antiandrogens and disorders of male sexual differentiation. *Molecular and Cellular Endocrinology* 178: 99–105

Tabb, M. M. and B. Blumberg. 2006. New modes of action for endocrine-disrupting chemicals. *Molecular Endocrinology* 20: 475–482

Taylor, C. A., M. H. Saint-Hilaire, L. A. Cupples, C. A. Thomas, A. E. Burchard, R. G. Feldman, and R. H. Myers. 1999. Environmental, medical, and family history risk factors for Parkinson's disease: a New England-based case control study. *American Journal of Medical Genetics* 88: 742–749

Turyk, M., H. Anderson, L. Knobeloch, P. Imm, and V. Persky. 2009. Organochlorine exposure and incidence of diabetes in a cohort of Great Lakes sport fish consumers. *Environmental Health Perspectives* 117: 1076–1082

Uemura, H., K. Arisawa, M. Hiyoshi, H. Satoh, Y. Sumiyoshi, K. Morinaga, K. Kodama, T. Suzuki, M. Nagai, and T. Suzuki. 2008. Associations of environmental exposure to dioxins with prevalent diabetes among general inhabitants of Japan. *Environmental Research* 108: 63–68

Vasiliu, O., L. Cameron, J. Gardiner, P. DeGuire, and W. Karmaus. 2006. Polybrominated biphenyls, polychlorinated biphenyls, body weight, and incidence of adult-onset diabetes mellitus. *Epidemiology* 17: 352–359

Weidner, I. S., H. Moller, T. K. Jensen, and N. E. Skakkebæk. 1998. Cryptorchidism and hypospadias in sons of gardeners and farmers. *Environmental Health Perspectives* 106: 793–796

World Health Organization. 2014. Suicide fact sheet. Details available at www. who.int/mediacentre/factsheets/fs398/en/

Wolff, M. S., J. A. Britton, S. L. Teitelbaum, S. Eng, E. Deych, K. Ireland, Z. Liu, A. I. Neugut, R. M. Santella, and M. D. Gammon. 2005. Improving organochlorine biomarker models for cancer research. *Cancer Epidemiology, Biomarkers and Prevention* 14: 2224–2236

Xiang, H., J. R. Nuckols, and L. Stallones. 2000. A geographic information assessment of birth weight and crop production patterns around mother's residence. *Environmental Research* 82: 160–167

Zorzon, M., L. Capus, A. Pellegrino, G. Cazzato, and R. Zivadinov. 2002. Familial and environmental risk factors in Parkinson's disease: a case-control study in north-east Italy. *Acta Neurologica Scandinavica* 105: 77–82

CHAPTER 5

EXPOSURE ASSESSMENT STUDIES

PESTICIDE EXPOSURE ASSESSMENT

Factors as varying as application intensity, method of application, frequency, behavioural characteristics of the applicator (for example, use of personal protective equipment and hand washing), as well as physical, chemical, and biological properties of the pesticide formulation (Hoppin, Adgate, Eberhart, *et al.* 2006), influence temporal variability in exposure. The frequency and duration of exposure are the design considerations for evaluating the temporal patterns of exposure to pesticides (Hoppin, Adgate, Eberhart, *et al.* 2006). The pesticide DDT (dichlorodiphenyltrichloroethane) and its primary metabolite and breakdown product DDE (dichlorodiphenyldichloroethylene) are still found in house dust in the United States even after 30 years of its ban on use (Lewis, Roberts, Chuang, *et al.* 1995; Whitmore, Immerman, Camann, *et al.* 1994). The widely used modern pesticides, such as organophosphates, carbamates, and pyrethroids, are less persistent than older formulations (Hoppin, Adgate, Eberhart, *et al.* 2006). As these compounds degrade over time, it is important to measure exposures shortly after application (Hoppin, Adgate, Eberhart, *et al.* 2006).

Although available data are limited, it is likely that pesticides commonly used today degrade more quickly in outdoor environments compared to indoor environment because in the latter case the pesticides are protected from rain, sunlight, temperature extremes, microbial action, and other degradation processes (Butte and Heinzow 2002; Lewis, Fortune, Blanchard, *et al.* 2001; Simcox, Fenske, Wolz, *et al.* 1995). Therefore, pesticides in homes may represent a longer term potential source of exposure with relatively less temporal variability than outdoor applications (Hoppin, Adgate, Eberhart, *et al.* 2006).

HUMAN HEALTH RISK ASSESSMENT

The US Environmental Protection Agency (EPA) evaluates the human health risk by following the National Research Council's four-step process (USEPA 2007), which is discussed in the following sections:

Step 1   Hazard identification (toxicology)

In the risk assessment process, the first step is to identify potential health effects associated with different types of pesticide exposure (USEPA 2007). Generally, for human health risk assessments, pesticide companies get many toxicity tests conducted on animals in independent laboratories which are then evaluated for acceptability by EPA scientists (USEPA 2007). EPA evaluates pesticides for a number of potential adverse effects, ranging from eye and skin irritation to cancer and birth defects in laboratory animals (USEPA 2007). EPA may also consult the public literature or other sources of supporting information on any aspect of the chemical (USEPA 2007).

Step 2   Dose–response assessment

The "father" of modern toxicology, the Swiss physician and alchemist, Paracelsus (1493–1541) had said: "The dose makes the poison." Both the amount of a substance to which a person is exposed and the toxicity of the chemical are equally important (USEPA 2007). For example, aspirin taken in small doses can be beneficial to people, but it can be fatal at higher doses. In some individuals, aspirin may be deadly even at very low doses (USEPA 2007).

Dose–response assessment is the determination of the dose levels that cause adverse effects in test animals and the use of these dose levels to calculate an equal dose in humans (USEPA 2007).

Step 3   Exposure assessment

There are three major routes for environmental and occupational exposure to pesticides: (i) inhaling pesticides (inhalation exposure), (ii) absorbing pesticides through the skin (dermal exposure), and (iii) getting pesticides in the mouth or digestive tract (oral exposure) (USEPA 2007). Pesticides can enter the body via any one or all the three routes depending on the situation (USEPA 2007). Individuals are exposed to pesticides through various sources, such as food, home and personal use products, and drinking water, and exposure can also be work related (see Table 1).

Table 1 Source of pesticides

Source	Description
Food	Most of the foods we eat are grown by using pesticides. Therefore, pesticide residues may be present inside or on the surfaces of these foods
Home and personal use pesticides	A number of pesticides are available which can be used in and around home to control insects, weeds, mould, mildew, bacteria, lawn and garden pests and to protect pets from pests such as fleas. Pesticides may also be used as insect repellents, which are directly applied to the skin or clothing
Pesticides in drinking water	Some pesticides applied to farmland or other land areas can find their way in small amounts to the groundwater or surface water system which can enter drinking water supplies
Worker-related exposure to pesticides	Pesticide applicators, vegetable and fruit pickers, and others are exposed to pesticides owing to the nature of their jobs. To overcome the hazards workers face from occupational exposure, EPA evaluates occupational exposure through a separate programme. All pesticides registered by EPA have been shown to be safe when they are properly used

Source USEPA (2007)

Step 4 Risk characterization

Risk characterization is the final step in assessing risks to human health from pesticide exposure (USEPA 2007). It is the process that combines the hazards, the extent of exposure to hazards, and the dose–response relationship to describe the overall risk from a pesticide (USEPA 2007). It explains the assumptions underlying the assessment of exposure as well as the uncertainties involved in the assessment of dose–response relation (USEPA 2007). The strength of the overall database is considered and conclusions are made (USEPA 2007). EPA evaluates both toxicity and exposure and assesses the risk associated with the pesticide use (USEPA 2007). Hence,

Risk = Toxicity × Exposure

It can be noted from the above equation that human health risk from pesticide exposure depends on both the toxicity of the pesticide and the likelihood of people coming into contact with it (USEPA 2007). At least some amount of both exposure and toxicity is necessary to result in a risk (USEPA 2007). For instance, if a pesticide is poisonous, but people are not exposed to it, then there is no risk. However, the use of pesticides is usually always associated with some toxicity and exposure, which results in a potential risk (USEPA 2007). EPA recognizes that toxic effects of a pesticide vary between different animal species and from person to person (USEPA 2007). To account for this variability, uncertainty factors are built into the risk assessment (USEPA 2007). These uncertainty factors result in an additional margin of safety for protecting people who may be exposed to pesticides (USEPA 2007).

Types of Toxicity Tests EPA Recommends for Human Health Risk Assessments

Environmental Protection Agency evaluates studies conducted over different periods of time and measures specific types of health effects (USEPA 2007). These tests are evaluated to identify the potential health impacts in infants, children, and adults (USEPA 2007).

(i) **Acute testing:** Short-term exposure—a single exposure (dose)
   - Oral, dermal (skin), and inhalation exposure
   - Neurotoxicity
   - Skin irritation
   - Skin sensitization
   - Eye irritation

(ii) **Sub-chronic testing:** Intermediate exposure—repeated exposure over a longer period of time (30–90 days) (USEPA 2007).
   - Oral, dermal (skin), and inhalation
   - Neurotoxicity (nervous system damage)

(iii) **Chronic toxicity testing:** Long-term exposure—repeated exposure that lasts for most of the test animal's lifespan (USEPA 2007). The purpose is to assess the effects of a pesticide after prolonged and repeated exposures (USEPA 2007).
   - Chronic effects (non-cancer)
   - Carcinogenicity (cancer)

(iv) **Developmental and reproductive testing:** Identifies effects in the foetus of an exposed pregnant female (birth defects) and determines how pesticide exposure affects the ability of a test animal to successfully reproduce (USEPA 2007).

(v) **Mutagenicity testing:** Assesses a pesticide's potential to affect the genetic components of a cell (USEPA 2007).

(vi) **Hormone disruption:** Measures the potential of pesticides to disrupt the endocrine system (USEPA 2007). The endocrine system is made up of a set of glands that secrete hormones which regulate the development, growth, reproduction, and behaviour of animals, including humans (USEPA 2007).

Some other subclinical tests for pesticide exposure are discussed next.

1   BLOOD CHOLINESTERASE ACTIVITY

1.1 Principle

The principle of the method is the measurement of the production rate of thiocholine as acetylthiocholine/butyrylthiocholinesterse is hydrolysed. This is accomplished by the continuous reaction of thiol with 5,5′-dithiobis (2-nitrobenzoate). The intensity of colour can be measured at 412 nm in a spectrophotometer. The chemicals involved are acetylthiocholine iodide, butyrylthiocholine iodide, Tris-HCl, SDS (sodium dodecyl sulphate), and DTNB (5,5′-dithiobis(2-nitrobenzoic acid)).

1.2   Preparation of Reagents

- **Tris-HCl buffer (pH 7.4):** 7.9 g of Tris-HCl and 0.145 g of acetylthiocholine iodide are dissolved in about 480 ml of distilled water and pH is adjusted to 7.4. Then the volume is made up to 500 ml with further double distilled water.
- **Tris-HCl diluting buffer (pH 7.4):** 7.9 g of Tris-HCl is dissolved in about 480 ml of double distilled water and pH is adjusted to 7.4. The volume is made up to 500 ml with double distilled water.
- **5, 5′-Dithiobis(2-nitrobenzoic acid) and sodium dodecyl sulphate solution:** 0.1 g of DTNB and 1.10 g of SDS were dissolved in 250 ml of Tris-HCl dilution buffer (which yields final concentration of 0.04% and 0.44%, respectively, in the reaction mixture).

1.3   Blood Cholinesterase Activity Measurements

The activity of cholinesterase (AChE and BChE) in blood was determined by the spectrophotometric method developed by Ellman, Curtney, Andrews, *et al.* (1961) and modified by Chambers and Chambers (1989). The activities of AChE and BChE were measured by taking acetylthiocholine iodide and butyrylthiocholine iodide, respectively, as substrates of the enzymatic reaction. For AChE, 4 ml total volume of incubating mixture consisting of 0.1 M Tris-HCl buffer (pH 7.4) containing 1.0 M acetylthiocholine iodide and 25 µl of 25 times diluted blood was taken. In the case of BChE, 0.1 M Tris-HCl buffer (pH 7.4) containing 1.0 mM butyrylthiocholine iodide and 25 µl of 20 times diluted blood was taken. This mixture for both AChE and BChE was incubated for 15 min at 37°C. The reaction was stopped after 15 min of incubation with a 0.5 ml mixture of DTNB and SDS. The absorbance was finally read at 412 nm and the activity was expressed as mmols hydrolysed/h/L blood (IU/L) (Ellman, Curtney, Andrews, *et al.* 1961 and modified by Chambers and Chambers 1989).

1.4   Calculation

$$\frac{IU}{L} = \frac{OD \text{ of test} \times 4 \times 1000 \times 25 \times 1000 \times 60}{15 \times 13{,}600 \times 1000 \times 1000 \times 0.025}$$

where

> 4 is the total volume of solution used in the reaction
>
> 15 is the incubating time
>
> 1000 is for converting molars into mM
>
> 13,600 is the extinction coefficient in moles
>
> 1000 is for converting mM
>
> 1000 is for converting millilitre to litre
>
> 25 is for blood dilution
>
> 60 is for converting minutes into hours
>
> 0.025 is for the total volume of blood taken
>
> IU/L is mmols hydrolysed/h/L of blood

2   PESTICIDE RESIDUES ANALYSIS IN BLOOD

2.1   Chemicals and Reagents

The standards of organochlorines such as α-HCH, β-HCH, γ-HCH, δ-HCH, *p,p*-DDE, *o,p*-DDT, *p,p*-DDD, and *p,p*-DDT were from Supelco. The organic solvents for pesticide determinations were of HPLC grade obtained from Merck, Germany. All the other chemicals used were of the highest purity available from commercial sources.

2.2   Extraction of Organochlorine Pesticides

Organochlorine pesticides were extracted based on the procedure suggested by El-Salem, Marei, Ruzo, *et al.* (1982). Briefly, 1 ml blood was vertex corrected with 3 ml *n*-hexane in a test tube for 2 min on vertex and the upper organic layer was transferred to a new test tube. This process is repeated two more times with 3 ml and 2 ml of *n*-hexane, respectively. Finally, the collected organic layer was concentrated to nearly 1 ml, and this concentrated extract of blood (1 ml) was passed through a column packed with activated florisil (1 inch) and anhydrous sodium sulphate. The column was pre-washed with 10 ml *n*-hexane. Organochlorines were eluted with 10 ml of 15% ether in *n*-hexane as an eluting solvent. The eluent was concentrated to 1 ml under a stream of nitrogen. An aliquot of 5 µl was injected into a gas liquid chromatograph (GLC) instrument with ^{63}Ni electron capture detector (ECD). To maintain accuracy, organochlorine standards were run every day on GLC–ECD before sample analysis was carried out. Sample peaks were identified by comparing their retention times with those of the standards. The quantitative concentration of pesticide isomers in the samples was calculated by the peak area of isomers and compared with that of the peak area of a known concentration of the respective isomers in an organochlorine standard chromatograph.

2.3 Gas Liquid Chromatography (Nucon-5743) Conditions for Pesticide Residue Analysis

Column—6 ft × 1.8 ft (i.d.) glass column with temperature 190°C; detector—^{63}Ni-ECD with temperature 250°C; injector temperature—250°C; carrier gas—IOLAR grade I nitrogen; flow rate—60 ml/min. The recovery experiment for pesticides was performed in triplicate and found to be 80%–90%. The level of detection (LOD) and level of quantification (LOQ) of chlorinated pesticide were 0.01 ppb and 1 ppb, respectively.

2.4 Pesticide Residue Concentration Calculation

$$\text{Concentration} \left(\mu g/ml\right) = \frac{\text{Area of unkown} \times \text{Amount injected (ng)} \times \text{Volume makeup (ml)}}{\text{Area of known} \times \text{Volume of blood (ml)} \times \text{Volume injected on GLC (µl)}}$$

2.5 Determination of Urinary Dialkylphosphate Metabolites of Organophosphate Pesticides Among Study Subjects

An analytical procedure described by Ueyama, Saito, Kamijima, *et al.* (2006) was standardized for the simultaneous determination of urinary dialkylphosphates (DAPs), metabolites of organophosphate pesticides, including diethylphosphate (DEP), dimethylphosphate (DMP), and dimethylthiophosphate (DMTP), in human urine samples, using a pentafluorobenzylbromide (PFBBr) derivatization and gas chromatography–mass spectrometry (GC–MS). The process of derivatization and detection of metabolites in urine is shown in Figure 1.

Pentafluorobenzylbromide-dialkylphosphate (PFB-DAP) was identified in the urine samples by GC–MS on the basis of C-ion (selected ions for confirmation) and Q-ion (selected ions for quantification). The retention times of PFB-DAPs were 14.17 min for DMP, 16.26 min for DEP, and 18.38 min for DETP. DMP metabolite in urine sample was selected on the basis of C-ions (306,194) and Q-ions (110). DEP was selected on the basis of C-ions (334,197) and Q-ions (258). DETP was selected on the basis of C-ions (350,213) and Q-ions (274).

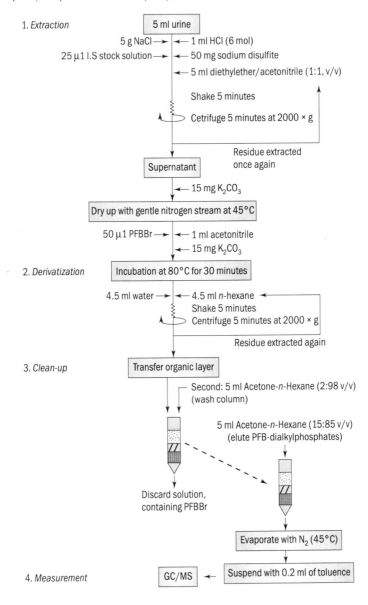

Figure 1 Analytical procedure for derivatization and detection of metabolites in urine

Source Ueyama, Saito, Kamijima, *et al.* (2006)

3 LUNG FUNCTION TEST

Peak expiratory flow rate (PEFR) and forced expiratory volume in 1 s (FEV_1) were measured using a spirometer (PIKO, UK) according to the recommendation of the American Thoracic Society (ATS) standards. The test was conducted under the supervision of a physiologist. The best values for PEFR and FEV_1 from three tests for each subject were recorded. The test was conducted as per the recommendations of ATS (Miller, Hankinson, Brusasco, *et al.* 2005).

4 NERVE CONDUCTION VELOCITY MEASUREMENTS

Nerve conduction studies were performed on EMG-EP Mark II (RMS, India). The parameters considered are as follows:

- Motor nerve conduction velocity (MNCV) study for median nerve
- Sensory nerve conduction study for median nerve

4.1 Placement of Electrodes (Mishra and Kalita 2004)

Stimulation point for median nerve (motor)

- Stimulator placed at wrist between Palmaris longus and flexor carpi radialis
- Ground electrode—dorsum of hand
- Recording electrode—active at the valley of abductor pollicis brevis
- Reference electrode—tendon of abductor pollicis brevis

Stimulation points for median nerve (sensory)

- Anti-dromic surface stimulation is performed at the wrist between palmarislongus and flexor carpi radialis tendons at second distal-most crease.
- Ground electrode—dorsum of hand
- Recording electrode—ring electrodes are used. Active electrode at the proximal inter phalangeal joint of second digit.
- Reference electrode—around distal phalynx of the same digit

Settings of RMS EMG machine for recording of motor conduction velocity and sensory conduction velocity of median nerve are given in Table 2.

Table 2 Settings of RMS EMG machine for recording of nerve conduction velocity of median nerve

Instrument setting	For recording MNCV	For recording SNCV (sensory nerve conduction velocity)
Sensitivity	2 mV	50 mV
Low frequency filter	2 Hz	2 Hz
High frequency filter	10 kHz	5 kHz
Sweep speed	5 ms/duration	5 ms/duration
Mode	Single	Single
Duration	100 μs	100 μs
Control	Remote	Remote
Range	100 mA	25 mA
Count	Infinite	Infinite
Pause	0	0
Rate	1 Hz	1 Hz

Source Sharma (2010)

Latent period was measured as time interval between stimulus artefact and onset of electrical response. Nerve conduction velocity (NCV) was expressed as nerve length divided by latent period. A metal tape was used to measure the nerve length. Sensory conduction velocity was determined by giving stimulus at the base of middle finger. Under the supervision of a physiologist, the electrophysiological NCV tests would be conducted on the arm using a standard protocol described by Kimura (1983). Lower limit of NCV normative data proposed by Kimura (1983) was used to identify study subjects with reduced motor and sensory NCV.

EMERGING ANALYTICAL TOOLS

There are emerging analytical tools for pesticides in different stages of research and development. Among them, a unique tool is the wearable sensor for pesticide exposure (Figure 2).

The emerging tools employed in projects on pesticide detection funded by National Institute of Environmental Health (NIEHS) are given in Table 3.

PESTICIDE EXPOSURE MONITORING METHODS

The two primary methods employed for evaluating exposure to pesticides are passive dosimetry, which is the more commonly used method, and

Figure 2 Wearable sensor for pesticide detection

Source Balshaw (2011)

Table 3 Emerging tools for pesticide detection

Title of the programme	Investigating group
HerbiScreen biosensor based on electrochemical detection of inhibition of photosynthesis of model algae	BioDetection Instruments LLC, Xiaoli Su
Disposable sensors based on SERS detection of organophosphates and organochlorines vapour sensors in air (vapour phase) and urine (aqueous phase)	EIC LABORATORIES INC., Kevin Spencer
Sensor based on carbon nanotubes for electrical impedance detection of pesticides in surface water	Dahl Natural, Anne Schwartz
Lateral flow-based assay for organophosphates (dialkylphosphates) in urine using high affinity insect chemosensory proteins	Inscent, Inc. Ken Konrad
GC-ECD method for detecting pesticides (organophosphates, organochlorines, and pyrethroids) in breast milk, baby food, and infant formula	Emory University, Barry Ryan

Source Balshaw (2011)

biological monitoring. Passive dosimetry measures the quantity of a pesticide that comes into contact with the skin, clothing, and the breathing zone of the worker. From the late 1960s until the 1980s, a large number of studies were performed on pesticide exposure in workers using passive dosimetry methods that measured the amount of pesticides coming into contact with the skin of a worker and the exposure was measured using patches attached to the clothing or from extracts of the clothing. The amount of pesticide that was absorbed into the body was not measured. These studies were reviewed by Wolfe (1976) and Davis (1980). For further details, refer to OECD (1997) and USEPA (1997).

Passive Dosimetry

Estimation of respiratory exposure

Air sampling for occupational exposure to pesticides involves measuring concentrations of pesticides in the breathing zone of workers with a portable air sampling pump and a sampling train. Sampling media are selected based on the physicochemical properties of the compound to be measured. Membrane filters, sorbent tubes, polyurethane foam, and charcoal are some examples of sampling media. Field workers may be exposed to chemical vapours, solid particulates, or water-based aerosols. For further details on the methods for respiratory exposure measurement, refer to Nigg, Beier, Carter, *et al.* (1990).

Estimation of dermal exposure

Dermal exposure sampling methods fall into three general categories: surrogate skin techniques, chemical removal techniques, and fluorescent tracer techniques (Fenske 1993a).

Surrogate skin techniques

Patches: Surrogate skin methods involve placing a collection medium against the skin or clothing and then analysing the chemicals deposited on it. The patch technique was developed by U.S. Public Health researchers in Wenatchee, Washington in response to potential acute intoxications among pesticide handlers (Durham and Wolfe 1962). This technique is now the internationally accepted standard (WHO 1986; USEPA 1987). Usually 10 patches are attached to the worker's clothing or skin, which can be on chest, back, upper arms, forearms, thighs, and lower legs. Chemical loading on the patch (mass per unit area) is then extrapolated to the skin surface area of the appropriate anatomical region. Patch technique for quantitative exposure estimation assumes that there is uniform exposure in a portion of the body; however, this assumption becomes invalid under certain circumstances (Franklin, Fenske, Greenhalgh, *et al.* 1981; Fenske 1990). Notwithstanding these limitations, patch sampling is a simple and cost-effective method for hazard evaluation and control. In a study of pesticide applicators in Brazil, patches were used to demonstrate the reduction in exposure resulting from changes in application equipment (Machado, Neto, Matuo, *et al.* 1992). Patches were also used in the studies of greenhouses to assess the effects of ventilation (Methner and Fenske 1994a). Numerous investigators have quantified protective clothing penetration by placing patch samplers inside and outside of the fabric barrier (Gold, Leavitt, Holsclaw, *et al.* 1982; Nigg, Stamper, and Queen 1986; Keeble, Dupont, Doucette, *et al.* 1988; Fenske, Blacker, Hamburger, *et al.* 1990; Nigg, Stamper, Easter, *et al.* 1992). The composition and size of the patches are taken into account in the study of dermal exposure and they should be based on the characteristics of the pesticide and the exposure scenario. Papermaking pulp or alpha cellulose is used for sprays because of the ability to absorb materials without disintegrating. Preparative chromatography paper is also utilized for the purpose. Other materials include surgical gauze, clothing material, and blotter paper. Patches constructed from surgical gauze are considered appropriate for dry formulations, such as dusts or granules.

Body garments: For registration purposes, whole-body garments have been used as a standard approach to assess workers' exposure to pesticides (Chester, Sabapathy, and Woollen 1992; Teschke, Marion, Jin, *et al.* 1994). Generally, whole-body garments are long underwear garments or coveralls worn next to the skin with no protective layer. Thus, there is an increased chance for residues to penetrate garments and reach the skin, which can lead to an underestimation of exposure. "Tyvek" coveralls are impermeable and give protection against pesticide penetration. Therefore, there is no underestimation of exposure to the skin in the case of Tyvek coveralls. Garments typically cover the torso and limbs and not the head, neck, face, hands, and feet. Thus, this method has an advantage over the patch technique in that it does not require extrapolation to total surface area for the torso.

To sample specific anatomical regions, use of body garments, such as gloves, can also be made (Davis, Stevens, and Staff 1983; Fenske, Birnbaum, Methner, *et al.* 1989; Brouwer, Marquart, De Mik, *et al.* 1992). Some investigators have reported that garment samplers such as gloves can overestimate exposures (Davis, Stevens, and Staff 1983), whereas others have found that there is no significant difference between glove and handwash measurements over an extended sampling period (Fenske, Birnbaum, Methner, *et al.* 1989). No standard materials have been developed for whole-body garment sampling. To date, studies have used absorbent fabric, such as cotton or cotton/polyester, and no any standard material has been developed for whole-body garments. Inner garments may include T-shirts (upper torso), white cotton socks (feet), briefs (lower torso), and thermal underwear bottoms and tops (the whole body except hands, feet, and head). The major assumption behind these methods is that the patch or garment captures and retains a chemical similar to that of skin. However, this assumption is not validated systematically. In fact, patches and garments, which are used as dermal samplers, should be pre-tested for their ability to absorb and retain the chemical under study.

Chemical removal techniques

Chemicals deposited on the skin can be removed by washing or wiping and measured (Durham and Wolfe 1962; Davis 1980). Wash techniques are employed to assess only the exposure of hand, while wiping techniques can, in theory, be applied to other skin surfaces.

Washes: To collect handwash samples, several types of solutions or liquids can be used, such as various types of aqueous surfactant solutions, neat isopropanol, or ethanol. The selection of the rinse solvent should be based on the physicochemical properties of the pesticide, especially the octanol–water partition coefficient (K_{ow}). For water-soluble pesticides, aqueous solutions should be preferred, whereas organic solvents are ideal for highly water-insoluble chemicals. Distilled or deionized water can be used for preparing aqueous solutions. Surfactants such as Sur-Ten, Aerosol OT-75, Emcol 4500, and Nekal WT-27 have been used to prepare hand-rinse solutions at about 0.01% concentrations. A number of procedures have been employed to prepare hand-rinse samples, which lead to the possibility that results obtained from the studies are probably not comparable. In some procedures, test subjects place their hands in a bowl (2–3 L) filled with rinse solutions and rub their hands together in a washing motion. In other cases, the liquid is slowly poured over the hands and the test subjects wring their hands in a washing motion. Then the solution is collected in a wide-top container. The method described by Durham and Wolfe (1962) and Davis (1980) is considered a standard procedure. In this method, hands are put in a plastic bag containing the solvent and shaken vigorously for 30 s; this procedure is repeated with a new bag and solvent, as well as for each hand. The four bags are then pooled. This method has shown good reproducibility in laboratory experiments (Fenske and Lu 1994). Handwash sampling method is used in the cases that involve frequent washing of hands. A wash should be carried out before the start of the exposure monitoring in order to remove any pre-existing residues.

The disadvantage of handwash techniques is that they do not necessarily remove completely the chemicals deposited on the skin. However, so far no studies have been conducted to verify this technique. In one study, less than 50% amount of chlorpyrifos applied on skin was found to be removed when it was washed 1 min after skin contact and washing 1 h after contact could remove less than 25% (Fenske and Lu 1994). There was a decrease in removal efficiency for low skin loadings. These results indicate that data obtained from such methods may be highly variable and appropriate removal efficiency studies will be required for validation of the method and quality assurance.

Wipes: Skin-wipe methods have also been developed to determine pesticide applicator exposure of hands, face, and neck, but these methods have not yet been tested and validated. A study reported that wiping of hands can remove pesticides with relatively high efficiency (Geno, Camann, Harding, *et al.* 1996). However, in that case the wiping was done immediately after exposure, and the effects of neither skin residence time nor concentration on removal efficiency were determined. A recent study of agricultural re-entry workers which involved the estimation of skin exposure found sixfold lower exposure rate for wiping sampling than for handwash sampling under similar exposure conditions (Fenske, Simcox, Camp, *et al.* 1999). Although skin wiping appears to be a relatively simple and convenient technique, it does not appear to be an acceptable quantitative exposure assessment method considering the conflicting findings.

Fluorescent tracer techniques

A relatively new method for assessing skin exposure patterns is the visualization method which uses fluorescent tracers. In the 1980s, fluorescent whitening agents (FWAs) were first demonstrated to be useful substances for characterizing deposition of pesticides on skin (Franklin, Fenske, Greenhalgh, *et al.* 1981; Fenske, Leffingwell, and Spear 1985; Fenske, Wong, Leffingwell, *et al.* 1986). Qualitative tracer studies can provide valuable insights into skin deposition patterns, protective clothing performance, and work practices (Fenske 1988). Pesticide exposure assessment is carried out by introducing a fluorescent tracer into the production system and then evaluating workers in a dark area using long-wavelength ultraviolet light. The tracer compounds are not visible under normal lighting conditions. The use of fluorescent compounds can be coupled with video imaging measurement to produce exposure estimates over virtually the entire body (Fenske and Birnbaum 1997). This approach involves taking pre- and post-exposure images of skin surfaces under long-wavelength ultraviolet illumination, developing a standard curve relating dermal fluorescence to skin-deposited tracer, and chemical residue sampling to estimate the quantity of the tracer and the chemical substance of interest deposited on the skin. This method was used to assess the performance of chemical protective clothing during application of ethion with airblast equipment in citrus orchards and it demonstrated limitations in garment design (Fenske 1993b). The technique has also been adopted to assess pesticide exposure during greenhouse applications (Methner and Fenske 1994a, 1994b). This method has been adopted by many laboratories, or similar approaches have been developed by them (Roff 1994; Archibald, Solomon, and Stephenson 1995; Bierman, Brouwer, and van Hemmen 1998; Kross, Nicholson, and Ogilvie 1996). Fluorescent tracer techniques show promise for improving accuracy in assessing dermal exposures. However, there are several disadvantages with this approach that makes the application of this method technically challenging. They are as follows:

1. In this method, introduction of the tracer compound into the agricultural spray mix is required.

2. There must be correlation between pesticide deposition and deposition of the fluorescent compound for the production, such that the fluorescence can be considered a tracer of chemical deposition.

3. To ensure an accurate measurement of the tracer, range finding and quality assurance studies are required.

4. It is necessary to characterize the relative penetration of the pesticide and tracer when protective clothing is worn by workers.

Estimation of exposure to children at home

Recently, many studies have been conducted on children's pesticide exposures in agricultural communities (Loewenherz, Fenske, Simcox, *et al.* 1997; Lu, Fenske, Simcox, *et al.* 2000). The preliminary work was on concentration measurements in air and house dust, as well as on surfaces (Hsu, Camann, Schattenberg, *et al.* 1990; Ross, Fong, Thongsinthusak, *et al.* 1991; Roberts, Budd, Ruby, *et al.* 1992; Simcox, Fenske, Wolz, *et al.* 1995). Studies involving evaluation of children's exposure to pesticides used in residential settings have adopted the approach employed in agricultural re-entry monitoring. Exposure occurs after application and is the product of environmental concentrations and contact rate.

Biological Monitoring

Biological monitoring is the estimation of internal dose of a compound in human body. Although biological monitoring of pesticide exposures—determining pesticides or their residues in biological fluids—has been conducted since the early 1950s, only recently this aspect has become the subject of renewed interest. Not all pesticides are amenable to biological monitoring. Pesticides that are rapidly absorbed and do not sequester or metabolize significantly are usually considered ideal for biological monitoring. Moreover, pesticides for which a quantitative relationship between exposure and urinary metabolites can be established are also considered suitable. About 80%–90% of the applied dose should be excreted in the urine within 5 days. Ideally, the pharmacokinetic model should demonstrate sufficient excretion in 1–3 days. If the pharmacokinetics in humans are not well characterized, biological monitoring should not be performed and reliance on animal data is insufficient. Biomarkers of exposure will be dealt here, including monitoring of pesticide metabolites in urine and of parent compounds in saliva. Several studies have been performed on the biological monitoring of pesticides (Wang, Franklin, Honeycutt, *et al.* 1989; He 1993; Woollen 1993; ICPS 1996). For further details, refer to 1987 USEPA Guideline for Applicator Exposure Monitoring (USEPA 1987).

Urinary metabolite monitoring

A more accurate estimate of internal dose can be obtained by measuring pesticide metabolites in urine. This approach is particularly useful when there are multiple routes of exposure—oral as well as respiratory and dermal—such as in the case of pesticide-exposed workers. If the total urinary output is collected, until either there are no detectable residues or background levels are reached (usually 48–96 h), the levels can be used to determine the internal dose. In both animals and humans, for several pesticides a good correlation between the amount of pesticide applied to the skin and the urinary output was shown in various studies (Franklin, Greenhalgh, and Maibach 1983; Franklin, Muir, and Moody 1986; Popendorf and Franklin 1987). However, there are weaknesses with this approach. The pharmacokinetics of the pesticide need to be known in humans. Highly volatile pesticides are extensively metabolized to minor metabolites or sequestered and are unlikely to result in an accurate dose estimate. For a complete discussion of this approach, refer to Woollen (1993) and OECD (1997). Urine samples can be collected by spot sampling (such as end of shift or morning void) or by 24 h sampling.

The 24 h sampling is more interpretable, but it is often difficult to obtain from workers. Sample collection is relatively simple and non-invasive, but it involves issues of privacy and confidentiality. Laboratory analysis is generally complex and therefore expensive. New analytical methods, such as enzyme-linked immunosorbent assays, are simple and cost-effective. Although such methods are useful to indicate exposure, they are not suitable for quantifying exposure.

To account for hydration effects, sometimes urinary metabolite measurements from spot samples are adjusted by urinary creatinine concentration. However, in this technique, the limitation is the high inter- and intra-individual variation of creatinine (Alessio, Berlin, Dell'Orto, *et al.* 1985; Boeniger, Lowry, and Rosenberg 1993). Creatinine levels may also be determined by a colorimetric method known as Jaffe reaction (Boeniger, Lowry, and Rosenberg 1993) or a specific gravity method (Alessio, Berlin, Dell'Orto, *et al.* 1985). Most clinical laboratories can perform these two analyses at relatively low cost. A urine specimen with creatinine and specific gravity values highly low or high should be viewed as suspect and possibly disregarded in a field study.

Salivary Monitoring

Saliva has been used to estimate plasma levels in humans for a variety of analytes. Measurements of pesticides in saliva have great potential because of the sampling simplicity and the expected reliability of salivary concentration as an indicator of tissue availability. Saliva has recently been used to indicate exposure for a few environmental contaminants, including pesticides (Nigg and Wade 1992). Saliva may be suitable for monitoring carbaryl exposure. Carbaryl concentrations in saliva were found to parallel those in plasma after gavage administration in rats (Skalasky, Lane, and Borzelleca 1979). Salivary concentration of ethion was measured among pesticide applicators (Nigg, Stamper, and Mallory 1993). Following ethion spraying in applicators, elevated salivary ethion levels were observed in them compared to controls, and some correlation was observed between urinary metabolite levels and salivary ethion levels ($r = 0.55$), which led the authors to conclude that saliva could be used to confirm ethion exposure. More recently, a study was conducted to investigate the feasibility of the herbicide atrazine monitoring using saliva by employing an animal model (Lu, Anderson, and Fenske 1997a; Lu, Anderson, Morgan, *et al.* 1997b). Not only salivary concentrations of atrazine were found to be highly correlated to plasma concentrations under varying conditions, but also salivary levels represent the portion of atrazine (protein unbound) in plasma with toxicological significance (Lu, Anderson, Morgan, *et al.* 1998). Recently, this technique was employed in the study of atrazine-exposed pesticide applicators, and the levels measured in saliva corresponded with pesticide application activities (Denovan, Lu, Hines, *et al.* 2000).

Validation of Passive Dosimetry

Comparison of estimates obtained from dosimetry with those from biological monitoring is one method to determine the accuracy of passive dosimetry at estimating dose. Regulatory agencies receive studies conducted using both techniques.

Atrazine

Several studies that employed passive dosimetry, biological monitoring, or both were submitted by the registrant to assess exposure to workers in the United States corn belt. The details of these studies are found in the USEPA Revised Human Health Risk Assessment (USEPA 2002) and the USEPA Re-registration Eligibility Document (USEPA 2003) on atrazine. In the United States, atrazine is predominantly used in agriculture, and it is the workers involved in mixing, loading, and applying atrazine who are most exposed to this chemical contaminant. The passive dosimetry studies reported atrazine residues only in terms of the parent compound. The biological monitoring studies measured chlorotiazenes metabolites. The atrazine absorbed dose was "back-calculated" from the measured metabolites based on a human excretion study.

REFERENCES

ACGIH (American Conference of Governmental Industrial Hygienists). 1990. *Documentation of the Threshold Limit Values and Biological Exposure Indices.* Cincinnati, OH: ACGIH

Alessio, L., A. Berlin, A. Dell'Orto, F. Toffoletto, and I. Ghezzi. 1985. Reliability of urinary creatinine as a parameter used to adjust values of urinary biological indicators. *International Archives of Occupational and Environmental Health* 55: 99–106

Archibald, B. A., K. K. Solomon, and G. R. Stephenson. 1995. Estimation of pesticide exposure to greenhouse applicators using video imaging and other assessment techniques. *American Industrial Hygiene Association Journal* 56: 226–235

Balshaw, D. M. 2011. Emerging analytical tools to assess exposure and the personal environment. Details available at <http://epa.gov/pesticides/ppdc/ testing/2011/october/analytical-tools.pdf>

Bierman, E. P. B., D. H. Brouwer, and J. J. van Hemmen. 1998. Implementation and evaluation of the fluorescent tracer technique in greenhouse exposure studies. *Annals of Occupational Hygiene* 42: 467–475

Boeniger, M. F., L. K. Lowry, and J. Rosenberg. 1993. Interpretation of urine results used to assess chemical exposure with emphasis on creatinine adjustments: a review. *American Industrial Hygiene Association Journal* 54: 615–627

Bradway, D. E., T. M. Shafik, and E. N. Lores. 1977. Comparison of cholinesterase activity, residue levels, and urinary metabolite excretion of rats exposed to organophosphorus pesticides. *Journal of Agricultural and Food Chemistry* 25(6): 1353–1358

Brouwer, R., H. Marquart, G. De Mik, and J. J. van Hemmen. 1992. Risk assessment of dermal exposure of greenhouse workers to pesticides after re-entry. *Archives of Environmental Contamination and Toxicology* 23: 273–280

Butte, W. and B. Heinzow. 2002. Pollutants in house dust as indicators of indoor contamination. *Reviews of Environmental Contamination and Toxicology* 175: 1–46

Calleman, C. J., L. Ehrenberg, B. Jansson, S. Osterman-Golkar, D. Segerback, K. Svenson, and C. A. Wachtmeister. 1978. Monitoring and risk assessment by means of alkyl groups in hemoglobin in persons occupationally exposed to ethylene oxide. *Journal of Environmental Pathology and Toxicology* 2: 427–442

Chambers, J. E. and H. W. Chambers. 1989. An investigation of acetylcholinesterase inhibition and aging and choline acetyltransferase activity following a high level acute exposure to paraoxon. *Pesticide Biochemistry and Physiology* 33: 125–131

Chester, G. 1993. Evaluation of agricultural worker exposure to, and absorption of, pesticides. *Annals of Occupational Hygiene* 37: 509–523

Chester, G., N. N. Sabapathy, and B. H. Woollen. 1992. Exposure and health assessment during application of lambda-cyhalothrin for malaria vector control in Pakistan. *Bulletin of the World Health Organization* 70: 615–619

Davis, J. E. 1980. Minimizing occupational exposure to pesticides: personal monitoring. *Residue Reviews* 75: 33–50

Davies, J. E., H. F. Enos, A. Barquet, C. Morgade, and J. X. Danauskas. 1979. Developments in toxicology and environmental science: pesticide monitoring studies. The epidemiologic and toxicologic potential of urinary metabolites. In *Toxicology and Occupational Medicine*, edited by W. B. Deichman, pp. 369–380. New York: Elsevier/North-Holland

Davis, J. E., E. R. Stevens, and D. C. Staff. 1983. Potential exposure of apple thinners to azinphosmethyl and comparison of two methods for assessment of hand exposure. *Bulletin of Environmental Contamination and Toxicology* 31: 631–638

Denovan, L. A., C. Lu, C. J. Hines, and R. A. Fenske. 2000. Saliva biomonitoring of atrazine exposure among herbicide applicators. *International Archives of Occupational and Environmental Health* 73: 457–462

Dong, M. H., J. H. Ross, T. Thongsinthusak, and R. I. Krieger. 1996. Use of spot urine sample results in physiologically based pharmacokinetic modeling of absorbed malathion doses in humans. *Biomarkers for Agrochemicals and Toxic Substances*. Washington, DC: American Chemical Society

Drevenkar, V. B., B. Stengl, B. Tkalcevic, and Z. Vasilic. 1983. Occupational exposure control by simultaneous determination of *N*-methylcarbamates and organophosphorus pesticide residues in human urine. *International Journal of Environmental Analytical Chemistry* 14: 215–230

Durham, W. F. and H. R. Wolfe. 1962. Measurement of the exposure of workers to pesticides. *Bulletin of the World Health Organization* 26: 75–91

Ellman, G. L., K. D. Curtney, V. Andrews, and R. M. Featherstone. 1961. A new and rapid colorimetric determination of acetyl cholinesterase activity. *Biochemical Pharmacology* 7: 88–95

El-Salem, A., M. Marei, L. O. Ruzo, and J. E. Casida. 1982. Analysis and persistence of permethrin, cypermethrin, deltamethrin and fenvalerate in the fat and brain of treated rats. *Journal of Agricultural and Food Chemistry* 30: 558–562

Fenske, R. A. 1988. Visual scoring system for fluorescent tracer evaluation of dermal exposure to pesticides. *Bulletin of Environmental Contamination and Toxicology* 41: 727–736

Fenske, R. A. 1990. Nonuniform dermal deposition patterns during occupational exposure to pesticides. *Archives of Environmental Contamination and Toxicology* 19: 332–227

Fenske, R. A. 1993a. Dermal exposure assessment technique. *Annals of Occupational Hygiene* 37: 687–706

Fenske, R. A. 1993b. *Fluorescent Tracer Evaluation of Protective Clothing Performance*. Cincinnati, OH: US Environmental Protection Agency

Fenske, R. A. and C. Lu. 1994. Determination of handwash removal efficiency: incomplete removal of the pesticide chlorpyrifos from skin by standard handwash techniques. *American Industrial Hygiene Association Journal* 55: 425–432

Fenske, R. A. and S. G. Birnbaum. 1997. Second generation video imaging technique for assessing dermal exposure (VITAE System). *American Industrial Hygiene Association Journal* 58: 636–645

Fenske, R. A., A. M. Blacker, S. J. Hamburger, and G. S. Simon. 1990. Worker exposure and protective clothing performance during manual seed treatment with lindane. *Archives of Environmental Contamination and Toxicology* 19: 190–196

Fenske, R. A., J. T. Leffingwell, and R. C. Spear. 1985. Evaluation of fluorescent tracer methodology for dermal exposure assessment. In *Dermal Exposure Related to Pesticide Use*, edited by R. C. Honeycutt, G. Zweig, and N. N. Ragsdale, ACS

Symposium Series 273, pp. 377–393. Washington, DC: American Chemical Society

Fenske, R. A., N. J. Simcox, J. E. Camp, and C. J. Hines. 1999. Comparison of three methods for assessment of hand exposure to azinphosmethyl (guthion) during apple thinning. *Applied Occupational and Environmental Hygiene* 14: 618–623

Fenske, R. A., S. G. Birnbaum, M. M. Methner, and R. Soto. 1989. Methods for assessing fieldworker hand exposure to pesticides during peach harvesting. *Bulletin of Environmental Contamination and Toxicology* 43: 805–815

Fenske, R. A., S. M. Wong, J. T. Leffingwell, and R. C. Spear. 1986. A video imaging technique for assessing dermal exposure. II. Fluorescent tracer testing. *American Industrial Hygiene Association Journal* 47: 771–775

Franklin, C. A., N. I. Muir, and R. P. Moody. 1986. The use of biological monitoring in the estimation of exposure during the application of pesticides. *Toxicology Letters* 33: 127–136

Franklin, C. A., R. A. Fenske, R. Greenhalgh, L. Mathieu, H. V. Denley, J. T. Leffingwell, and R. C. Spear. 1981. Correlation of urinary pesticide metabolite excretion with estimated dermal contact in the course of occupational exposure to Guthion. *Journal of Toxicology and Environmental Health* 7: 715–731

Franklin, C. A., R. Greenhalgh, and H. I. Maibach. 1983. Correlation of urinary dialkyl phosphate metabolite levels with dermal exposure to azinphos-methyl. In *Human Welfare and the Environment*, edited by J. Miyamoto and P. C. Kearney, Vol. 3, pp. 221–226. Oxford, UK: Pergamon Press

Gallop, B. R. and W. I. Glass. 1979. Urinary arsenic levels in lumber treatment operators. *New Zealand Medical Journal* 89: 10–11

Geno, P. W., D. E. Camann, H. F. Harding, K. Villalobos, and R. G. Lewis. 1996. Handwipe sampling and analysis procedure for the measurement of dermal contact with pesticides. *Archives of Environmental Contamination and Toxicology* 30: 132–138

Gold, R. E., J. R. C. Leavitt, T. Holsclaw, and D. Tupy. 1982. Exposure of urban applicators to carbaryl. *Archives of Environmental Contamination and Toxicology* 11: 63–67

He, F. 1993. Biological monitoring of occupational pesticides exposure. *International Archives of Occupational and Environmental Health* 65: S68–S76

Hoppin, J. A, J. L. Adgate, M. Eberhart, M. Nishioka, and B. P. Ryan. 2006. Environmental exposure assessment of pesticides in farmworker homes. *Environmental Health Perspectives* 114(6): 929–935

Hsu, J. P., D. Camann, H. Schattenberg, H. Wheeler, K. Villalobos, M. Kyle, and S. Quarderer. 1990. New dermal exposing sampling technique. *Measurement of Toxic and Related Air Pollutants*, pp. 489–497. Pittsburgh, PA: Air and Waste Management Association

ICPS (International Centre for Pesticide Safety). 1996. *Biological Monitoring of Human Exposure to Pesticides*. Milan, Italy: ICPS (draft)

Keeble, V. B., R. R. Dupont, W. J. Doucette, and M. Norton. 1988. Guthion penetration of clothing materials during mixing and spraying in orchards. In *Performance of Protective Clothing: Second Symposium*, edited by S. Z. Mansdorf, R. Sagar, and A. P. Nielson, pp. 573–583. Philadelphia, PA: American Society for the Testing of Materials

Kimura, J. 1983. *Electrodiagnosis in Diseases of Nerve and Muscle: Principles and Practice*. Philadelphia, PA: FA Davis Company

Kolmoden-Hedman, B., S. Hoglund, A. Swenson, and M. Okerblom. 1983. Studies on phenoxy acid herbicides. II. Oral and dermal uptake and elimination in urine of MCPA in humans. *Archives of Environmental Contamination and Toxicology* 54: 267–273

Krieger, R. I., T. M. Dinoff, and J. Peterson. 1996. Human disodium octaborate tetrahydrate exposure following carpet flea treatment is not associated with significant dermal absorption. *Journal of Exposure Analysis and Environmental Epidemiology* 6(3): 279–288

Kross, B. C., H. F. Nicholson, and L. K. Ogilvie. 1996. Methods development study for measuring pesticide exposure to golf course workers using video imaging techniques. *Applied Occupational and Environmental Hygiene* 11: 1346–1350

Kutz, F. W. and S. C. Strassman. 1977. Human urinary metabolites of organophosphate insecticides following mosquito adulticiding. *Mosquito News* 37(12): 211–218

Levy, K. A., S. S. Brady, and C. D. Pfaffenberger. 1981. Chlorobenzilate residues in citrus worker urine. *Bulletin of Environmental Contamination and Toxicology* 27(2): 235–238

Lewis, R. G., J. W. Roberts, J. C. Chuang, D. E. Camann, and M. G. Ruby. 1995. Measuring and reducing exposure to the pollutants in house dust. *American Journal of Public Health* 85: 1168

Lewis, R. G., C. R. Fortune, F. T. Blanchard, and D. E. Camann. 2001. Movement and deposition of two organophosphorus pesticides within a residence after interior and exterior applications. *Journal of the Air and Waste Management Association* 51: 339–351

Lieben, J. R., K. Waldman, and L. Krause. 1953. Urinary excretion of paranitrophenol following exposure to parathion. *AMA Archives of Industrial Hygiene and Occupational Medicine* 7: 93–98

Loewenherz, C., R. A. Fenske, N. J. Simcox, G. Bellamy, and D. Kalman. 1997. Biological monitoring of organophosphorous pesticide exposure among children of agricultural workers. *Environmental Health Perspectives* 105: 1344–1353

Lu, C., L. C. Anderson, and R. A. Fenske. 1997a. Determination of atrazine levels in whole saliva and plasma in rats: potential of salivary monitoring for occupational exposure. *Journal of Toxicology and Environmental Health* 50: 101–111

Lu, C., L. C. Anderson, M. S. Morgan, and R. A. Fenske. 1997b. Correspondence of salivary and plasma concentrations of atrazine in rats under variable salivary flow rate and plasma concentration. *Journal of Toxicology and Environmental Health* 52: 317–329

Lu, C., L. C. Anderson, M. S. Morgan, and R. A. Fenske. 1998. Salivary concentration of atrazine can be used to determine free atrazine plasma levels in rats. *Journal of Toxicology and Environmental Health* 53: 283–292

Lu, C., R. A. Fenske, N. J. Simcox, and D. Kalman. 2000. Pesticide exposure of children in an agricultural community: evidence of household proximity to farmland and take home exposure pathways. *Environmental Research* 84: 290–302

Machado, J., G. Neto, T. Matuo, and Y. K. Matuo. 1992. Dermal exposure of pesticide applicators in staked tomato (*Lycopersicon esculentum* Mill) crops: efficiency of a safety measure in the application equipment. *Bulletin of Environmental Contamination and Toxicology* 48: 529–534

Methner, M. M. and R. A. Fenske. 1994a. Pesticide exposure during greenhouse applications, Part I. Dermal exposure reduction due to directional ventilation and worker training. *Applied Occupational and Environmental Hygiene* 9: 560–566

Methner, M. M. and R. A. Fenske. 1994b. Pesticide exposure during greenhouse applications, Part II. Chemical permeation through protective clothing in contact with treated foliage. *Applied Occupational and Environmental Hygiene* 9: 567–574

Miller, M. R., J. Hankinson, V. Brusasco, F. Burgos, R. Casaburi, A. Coates, R. Crapo, P. Enright, C. P. van der Grinten, P. Gustafsson, R. Jensen, D. C. Johnson, N. MacIntyre, R. McKay, D. Navajas, O. F. Pedersen, R. Pellegrino, G. Viegi, and J. Wanger. 2005. Standardisation of spirometry. *European Respiratory Journal* 26(2): 319–338

Mishra, U. K. and J. Kalita (eds). 2004. *Clinical Neurology: nerve conduction electormyography and evoked potentials,* 2nd edn. New Delhi: Elsevier

Neumann, H. 1984. Analysis of hemoglobin as a dose monitor for alkylating and arylating agents. *Archives of Toxicology* 56: 1–6

Nigg, H. N. and S. E. Wade. 1992. Saliva as a monitoring medium for chemicals. *Reviews of Environmental Contamination and Toxicology* 129: 95–119

Nigg, H. N., H. H. Stamper, and R. M. Queen. 1986. Dicofol exposure to Florida citrus applicators: effects of protective clothing. *Archives of Environmental Contamination and Toxicology* 15: 121–134

Nigg, H. N., R. C. Beier, O. Carter, C. Chaisson, C. Franklin, T. Lavy, R. G. Lewis, P. Lombardo, J. F. McCarthy, K. T. Maddy, M. Moses, D. Norris, C. Peck, K. Skinner, and R. G. Tardiff. 1990. Exposure to pesticides. In *The*

Effect of Pesticides on Human Health, edited by S. R. Baker and C. F. Wilkinson, pp. 35–130. Princeton, NJ: Princeton Scientific Publishing

Nigg, H. N., J. H. Stamper, E. Easter, and J. O. Dejong. 1992. Field evaluation of coverall fabrics: heat stress and pesticide penetration. *Archives of Environmental Contamination and Toxicology* 23: 281–288

Nigg, H. N., J. H. Stamper, and L. L. Mallory. 1993. Quantification of human exposure to ethion using saliva. *Chemosphere* 26: 897–906

NIOSH (National Institute of Occupational Safety and Health). 1984. *NIOSH Manual of Analytical Methods*. Cincinnati, OH: NIOSH

OECD. 1997. Guidance document for the conduct of studies of occupational exposure to pesticides during agricultural application. Series on Testing and Assessment No. 9, OCDE/GD(97)148. Organization for Economic Co-operation and Development, Paris

Peoples, S. A. and J. B. Knaak. 1982. Monitoring pesticide safety programs by measuring blood cholinesterase and analyzing blood and urine for pesticides and their metabolites. In *Pesticide Residue and Exposure*, edited by J. R. Plimmer, ACS Symposium Series 182, pp. 41–57. Washington, DC: American Chemical Society

Pereira, M. A. and L. W. Chang. 1982. Hemoglobin binding as a dose monitor for chemical carcinogens. In *Bradway Report 13, Indicators of Genotoxic Exposure*, edited by B. A. Bridges, B. E. Butterworth, and I. B. Weinstein, pp. 177–187. Cold Spring Harbor, NY: Cold Spring Harbor Laboratory

Popendorf, W. 1992. Reentry field data and conclusions. *Reviews of Environmental Contamination and Toxicology* 128: 71–117

Popendorf, W. and C. A. Franklin. 1987. Pesticide exposure assessment. In *Pesticide Science and Biotechnology*, edited by R. Greenhalgh and T. P. Roberts, pp. 565–568. Oxford, UK: Blackwell Scientific Publishers

Popendorf, W. and J. T. Leffingwell. 1982. Regulating organophosphate pesticide residues for farmworker protection. *Residue Reviews* 85: 125–201

Ritter, L. and C. A. Franklin. 1989. Use of biological monitoring in the regulatory process. In *Biological Monitoring for Pesticide Exposure*, ACS Symposium Series 382. Washington, DC: American Chemical Society

Roan, C. C., D. P. Morgan, N. Cook, and E. H. Paschal. 1969. Blood cholinesterases, serum parathion concentrations and urine *p*-nitrophenol concentrations in exposed individuals. *Bulletin of Environmental Contamination and Toxicology* 4: 362–369

Roberts, J. W., W. T. Budd, M. G. Ruby, D. E. Camann, R. C. Fortmann, R. G. Lewis, L. A. Wallace, and T. M. Spittler. 1992. Human exposure to pollutants in the floor dust of homes and offices. *Journal of Exposure Analysis and Environmental Epidemiology* 2: 127–146

Roff, M. W. 1994. A novel lighting system for the measurement of dermal exposure using a fluorescent dye and an image processor. *Annals of Occupational Hygiene* 38: 903–919

Ross, J., H. R. Fong, T. Thongsinthusak, S. Margetich, and R. Krieger. 1991. Measuring potential dermal transfer of surface pesticide residue generated from indoor fogger use: using the CDFA roller method. Interim report II. *Chemosphere* 22: 975–984

Sharma, G. 2010. Detection of median nerve entrapment in the carpal tunnel by electrodiagnostic techniques. *Calicut Medical Journal* 8(1): e3

Simcox, N. J., R. A. Fenske, S. A. Wolz, I. C. Lee, and D. A. Kalman. 1995. Pesticides in household dust and soil: exposure pathways for children of agricultural families. *Environmental Health Perspectives* 103: 1126–1134

Skalasky, H. L., R. W. Lane, and J. F. Borzelleca. 1979. Excretion of carbaryl into saliva of the rat and its effect on cholinesterase. In *Toxicology and Occupational Medicine*, edited by W. B. E. Reichmann, pp. 22–25. Miami, FL: Elsevier Science

Swan, A. A. B. 1969. Exposure of spray operators to paraquat. *British Journal of Industrial Medicine* 26: 322–329

Tannenbaum, S. R. and P. L. Skipper. 1984. Biological aspects to the evaluation of risk: dosimetry of carcinogens in man. *Fundamental and Applied Toxicology* 4: S367–S373

Teschke, K., S. A. Marion, A. Jin, R. A. Fenske, and C. van Netten. 1994. Strategies for determining occupational exposures in risk assessments: a review and a proposal for assessing fungicide exposures in the lumber industry. *American Industrial Hygiene Association Journal* 55: 443–449

Ueyama, J., I. Saito, M. Kamijima, T. Nakajima, M. Gotoh, T. Suzuky, E. Shibata, T. Kondo, K. Takagi, K. Miyamoto, J. Takamatsu, T. Hasegawa, K. Takagi. 2006. Simultaneous determination of urinary dialkylphosphate metabolites of organophosphorus pesticides using gas chromatography–mass spectrometry. *Journal of Chromatography B: Analytical Technologies in the Biomedical and Life Sciences* 832: 58–66

USEPA (US Environmental Protection Agency). 1987. *Pesticide Assessment Guidelines, Subdivision U, Applicator Exposure Monitoring, Report No. 540/9-87-127.* Washington, DC: Office of Prevention, Pesticides and Toxic Substances, USEPA

USEPA (US Environmental Protection Agency). 1997. *Standard Operating Procedures (SOPs) for Residential Exposure Assessments: draft report.* Washington, DC: USEPA

USEPA (US Environmental Protection Agency). 2002. *Atrazine Revised Human Health Assessment for the Reregistration Eligibility Document (RED),* Case No. 0062. Washington, DC: USEPA

USEPA (US Environmental Protection Agency). 2003. *Interim Reregistration Eligibility Document for Atrazine,* Case No. 0062. Washington, DC: USEPA

USEPA (US Environmental Protection Agency). 2007. Assessing health risks from pesticides. Details available at <www.epa.gov/opp00001/factsheets/riskassess. htm>

Wagner, S. L. and P. Weswig. 1974. Arsenic in blood and urine of forest workers. *Archives of Environmental Health* 28: 77–79

Wang, R. G., C. A. Franklin, R. C. Honeycutt, and J. C. Reinert (eds). 1989. *Biological Monitoring for Pesticide Exposure,* ACS Symposium Series 382. Washington, DC: American Chemical Society

Whitmore, R. W., F. W. Immerman, D. E. Camann, A. E. Bond, R. G. Lewis, and J. L. Schaum. 1994. Nonoccupational exposures to pesticides for residents of two U.S. cities. *Archives of Environmental Contamination and Toxicology* 26: 47–59

WHO (World Health Organization). 1986. Field surveys of exposure to pesticides: standards protocol. *Toxicology Letters* 33: 223–235

Wilson, H. K. 1986. Breath analysis: physiological basis and sampling techniques. *Scandinavian Journal of Work, Environment and Health* 12: 174–192

Woollen, B. H. 1993. Biological monitoring for pesticide absorption. *Annals of Occupational Hygiene* 37: 525–540

CHAPTER 6

HUMAN STUDIES IN DEVELOPING COUNTRIES

ESTIMATES OF PESTICIDE PRODUCTION AND PESTICIDE POISONING

In developing nations, extensive use of agricultural pesticides without taking enough precaution is causing a lot of health concerns. At present, acute poisoning owing to exposure to agricultural pesticides is largely responsible for the increased cases of human morbidity and mortality worldwide. Around 25 million farm workers are exposed to pesticides in the developing countries (Jeyaratnam 1990). Although developing countries account for only 20% of the agrochemicals used in the world, 99% of deaths from pesticide poisoning occur in these countries (Jeyaratnam and Chia 1994). According to an estimate of the World Health Organization (WHO), 3 million episodes of acute pesticide poisoning occur annually at the global level. Of these a minimum of 300,000 die, with 99% of the cases reported from low- and middle-income countries (Gunnell and Eddleston 2003). In Indonesia, 265 different pesticides are registered, and 44% of their active ingredients fall in the categories Ia, Ib, II (extremely, highly, and moderately hazardous, respectively) of the WHO hazard grades (Kishi, Hirschhorn, Djajadisastra, *et al.* 1995). India is the largest manufacturer of basic pesticides in Asia and ranks twelfth globally. Insecticides account for 75% of India's total pesticide consumption, followed by fungicides (12%) and herbicides (10%). Of the total quantity of pesticides used in India, over 50% are consumed in the cultivation of cotton, 17% in rice, and 13% in vegetables and fruits (Indra, Bellamy, and Shyamsundar 2007). In the early 2000s, the number of deaths caused by pesticides in Japan was about 1000 persons per year (Nagami, Nishigaki, Matsushima, *et al.* 2005). Currently, Japanese farmers use the estimated 400,000 tonnes of pesticides per year.

CONCERNS IN THE AGRARIAN SECTOR DUE TO PESTICIDE SPRAYING

Pesticides used in agricultural tracts are released into the atmosphere and they come into human contact either directly or indirectly. Pesticides present in environmental media (e.g. soil, water, air, and food) come into contact with human populations by different routes of exposure such as inhalation, ingestion, and dermal contact. Exposure to pesticides results in acute and chronic health problems. These problems range from temporary acute effects such as irritation of eyes and excessive salivation to chronic diseases such as cancer and reproductive and developmental disorders (Yassi, Kjellstrom, Kok, *et al.* 2001).

Currently, estimates of the hazards associated with long-term exposure to pesticides are not available. The necessity to ensure local agricultural production and food security in low- and middle-income countries and simultaneously protect the population against the health impacts following exposure to pesticides has emerged as a major global public health challenge. Acute pesticide poisoning is still not as pronounced in Africa as that in Asia. The increased agricultural production in Africa and widespread use of pesticides in agriculture will significantly increase the number of pesticide poisoning cases (London, Flisher, Wesseling, *et al.* 2005). The Poison Information Centre at the National Institute of Occupational Health in Ahmedabad reported that organophosphorus pesticides were responsible for the maximum number of poisonings (73%) among all agricultural chemicals (Dewan and Saiyed 1998).

In developing countries, acute and chronic poisoning among pesticide applicators is still a major health issue in rural areas, but there are few published reports. The spray practices followed by Indonesian farmers substantially exposed them to pesticides and they developed signs and symptoms significantly more often during spraying seasons than during non-spraying seasons (Kishi, Hirschhorn, Djajadisastra, *et al.* 1995). The number of spray operations per week, the use of hazardous pesticides, and skin and clothes wetted with the spray solution were significantly and independently associated with the number of signs and symptoms related to neurobehavioural, respiratory, and intestinal problems in Indonesian farmers (Kishi, Hirschhorn, Djajadisastra, *et al.* 1995). In developing countries, there is continuous use of different pesticides, including banned ones, based on the season and crop produced. The application of pesticides also poses health risks to households using the pesticides without prior knowledge of their toxic effects and without following safety precautions (Jamil, Das, Shaik, *et al.* 2007).

PESTICIDE POISONING INCIDENTS AND FARM PRACTICES IN DEVELOPING COUNTRIES

In India, a large percentage of population (56.7%) is engaged in agriculture and, therefore, exposed to the pesticides used in agriculture (Gupta 2004; Government of India 2001). India ranks second in Asia in annual pesticide consumption (O'Malley 1997). The first case of pesticide poisoning in India was reported from Kerala in 1958, where more than 100 people died after consuming wheat flour contaminated with parathion (Karunakaran 1958). The chemical used was ethyl parathion, known as Folidol E-605. In 1977, an outbreak of food poisoning occurred in a village in Uttar Pradesh following accidental ingestion of hexachlorocyclohexane (HCH)-contaminated wheat in which eight cases of grand mal seizures were reported (Nag, Singh, and Senon 1977). The malathion (diazole) poisoning in Indore during 1967–68 claimed five lives, out of the 35 cases reported. In all the cases, electrocardiogram (ECG) alterations were recorded and autopsy and histopathological studies revealed damage to the myocardium (Sethuraman 1977). In India, the presence of pesticide residues is reported in 51% of food commodities and out of these, 20% have pesticide residues above the maximum residue level (Gupta 2004). It has been found that long-term, low-dose exposures to pesticide residues are associated with human health effects, such as immune suppression, diminished intelligence, hormone disruption, reproductive abnormalities, and cancer (Gupta 2004). Jamil, Das, Shaik, *et al.* (2007) found in their study that Indians in rural areas are more vulnerable to the pesticide exposure during agricultural practices. The incidental and occupational (agriculture) exposure of pesticides and the duration of exposure to pesticide poisoning in developing countries are given in Table 1.

In developing countries, pesticide poisoning is a major health problem which largely occurs owing to traditional methods followed in the application of pesticides. Farm workers do not follow appropriate safety measures while handling and spraying pesticides. Moreover, in addition to exposure during pesticide applications, workers are at risk of exposure during mixing and handling chemicals, during cleaning and loading spray equipment, and during disposal of empty containers. Another alarming concern in developing countries is the consumption of pesticides to commit suicide by farmers trapped in debt or facing other financial crises owing to crop loss. There are increasing reports of such incidents in countries such as Sri Lanka and India (Jeyaratnam 1990; Eddleston 2000; Eddleston and Phillips 2004). Such incidents occur mainly among males. More than 80%

Table 1 Incidental and occupational (agriculture) exposure to pesticides in developing countries

Types of pesticides	Effects and symptoms	Country; reference	Exposure time (year)	Types of poisoning (incidental or occupational)
DDT, DDE, lindane, arachlor, heptachlor, aldrin, endrin, HCH	Precocious puberty (endocrine disorder) among exposed children	Belgium (Konstantinova, Charlier, Craen, *et al.* 2001)	9	Incidental
Methamidophos, carbofuran, monocrotophos	Acute poisoning led to mortality of 12.98%	Brazil (Maria, Dario, and Eloisa 2006)	10	Occupational (agricultural)
2,4-D, picloram, glyphosate, benomyl, chlorothalonil, paraquat, carbofuran, propiconazole, mancozeb, terbufos, methamidophos, deltamethrin, methomyl, triadimefon, fluazifop, captafol, lead arsenate, fenamiphos, phoxim, malathion, dichlorvos, terbuthylazine, diuron, oxamyl, quintozene, aldrin	Childhood leukaemia	Costa Rica (Patricia, Partanen, Wesseling, *et al.* 2005)	21	Occupational (agriculture)
Chlordimeform, chlorpyrifos, diazinon, fenthion, sulphoxide, mephosfolan, methyl parathion, p,p´-DDE, $p,p´$-DDD, o,p-DDT, $p,p´$-DDT, β-HCH, lindane (γ-HCH), toxaphene	Decreased cholinesterase activity among exposed children	Nicaragua (McConnell, Pacheco, Wahlberg, *et al.* 1999)	—	Incidental

Contd...

Table 1 Contd...

Types of pesticides	Effects and symptoms	Country, reference	Exposure time (year)	Types of poisoning (incidental or occupational)
Organophosphate and carbamate pesticides	Headache, burning sensation in eyes/face, weakness, fever watering eyes, blurred vision, skin, irritation/itching, dizziness, nausea and vomiting, cold/breathlessness/ chest pain, forgetfulness male impotence, female infertility, decreased cholinesterase activity	Ghana (Clarke, Levy, Spurgeon, *et al.* 1997)	1–21	Occupational (agricultural)
Methyl bromide, methamidophos (Tameron®), DDVP, folimat, captan, folidol, bazudin, agrimec, others (pyrethroid, carbamate, chlorohydrocarbons)	Fatigue, headache, changes in mood, irritation in eyes, blurring of vision, difficulty breathing, pressure on chest, coughing, mucous saliva, skin itch, scars, dizziness, depression, forgetfulness, memory disorders, sleeplessness, nausea, vomiting, bleeding nose, weight loss, deformity of nails	Turkey (Ergonen, Salacin, and Ozdemir 2005)	14	Occupational (agricultural)
Cartap, isoprothiolane, fenobucarb, thiosultap-sodium, difenoconazole, sulphur, isoprothiolane, pretilachlor, butachlor, phthalide, iprodione, tricyclazole, carbendazim, acephate, diazinon, hexaconazole, 2,4-D quinalphos, carbofuran, validamycin, iprobenfos (iprobenphos), carbendazim, tricyclazole, imidacloprid, propiconazole, thiosultap-sodium (nereistoxin/dimehypo), fenitrothion, chlorpyrifos, phthalide glyphosate, zineb, fipronil, tricyclazole, mancozeb, permethrin, bispyribac-sodium, fenobucarb, endosulfan, cypermethrin, benomyl, butachlor	Skin irritation, headache, dizziness, eye irritation, shortness of breath, acetyl cholinesterase inhibition	Vietnam (Dasgupta, Meisner, Wheeler, *et al.* 2007)	9	Occupational (agricultural)

Contd...

Table 1 Contd...

Types of pesticides	Effects and symptoms	Country; reference	Exposure time (year)	Types of poisoning (incidental or occupational)
Methyl parathion, monocrotophos, acephate, malathion, phorate, triazophos, quinalphos, chlorpyrifos, endosulfan, endrin, indoxacarb, cypermethrin, spinosad, imidacloprid	Acute poisoning led to death of 22.6% subjects	India (Rao, Venkateswalru, Surender, et al. 2005)	6	Incidental
Endosulfan	Reduced male reproductive development among exposed children	India (Saiyed, Dewan, Bhatnagar, et al. 2003)	20	Occupational (agricultural)
Monocrotophos, phosphamidon, dichlorvos, oxydemeton methyl, malathion, endosulfan, methyl parathion, dimethoate or carbaryl	Respiratory problems, ocular problems, gastrointestinal problems, dermal problems, decreased acetyl cholinesterase activity, elevated MDA levels	India (Kesavachandran, Rastogi, and Mathur 2006a)	5	Occupational (agricultural)
Acephate, chlorpyriphos, chlordane, dimethoate, allethrin, pipernoyl butoxide, transfluthrin, endosulfan, chlorpyrifos, profenofos, monocrotophos, carbofuran, cypermethrin, cyfluthrin	Decrease in red blood cell count and haemoglobin, increase in white blood cell count with a large number of immature cells and diagnosed as Philadelphia negative chronic myeloid leukaemia, childhood signs of mental retardation	India (Jamil, Das, Shaik, et al. 2007)	1–5	Occupational (agricultural)
Organophosphates, including fenthion and some pyrethroids	Cytogenetic damage	Pakistan (Bhalli, Khan, Haq, et al. 2006)	13	Occupational (agricultural)

Contd...

Table 1 Contd...

Types of pesticides	Effects and symptoms	Country; reference	Exposure time (year)	Types of poisoning (incidental or occupational)
pp′-DDE, *pp*-DDD, *pp′*-DDT, *op*-DDT, and α,β,γ, δ-HCH	Inhibited acetyl- and butryl-cholinesterase activities and showed elevated MDA levels; respiratory morbidity, ocular problems, gastrointestinal and skin problems	India (Singh, Jyoti, Reddy, *et al.* 2007)	5	Occupational (agricultural)

DDD, dichlorodiphenyldichloroethane; DDE, dichlorodiphenyldichloroethylene; DDT, dichlorodiphenyltrichloroethane; DDVP, dichlorvos; HCH, hexachlorobenzene; MDA, malondialdehyde of organophosphate poisoning cases in Sri Lanka result from intentional oral intake with suicidal intent (Senanayake 1998). Jeyaratnam (1985) found that among all the cases of pesticide poisoning in Sri Lanka, 73% were suicidal, 17% occupational, and 8% accidental.

Less than 2% of the applicators understand the toxicity levels of the pesticides they use (Indra, Bellamy, and Shyamsundar 2007). Only about a third of the applicators read the information on the label before using a chemical, while only 2.5% of them follow the label instructions. Only 1.5% understand the toxicity level associated with the colour code system used in the region. Even sprayers who are aware of the potential health hazards associated with pesticides and the need for personal protection do not follow the recommended protective measures such as the use of face masks with replaceable filters and rubber gloves. Cost, general lethargy, and discomfort are the main reasons why such devices are not used. Some make-shift protection is used, such as shirtsleeves or a cloth wrapped around the nose; however, such measures offer little defence (Indra, Bellamy, and Shyamsundar 2007). In developing countries, the practice of chewing or smoking by sprayers "to reduce the nauseating feeling" while spraying pesticides also increases the magnitude of health risks. Another practice in developing countries is combining multiple pesticides, many of which have different trade names but the same common name and so the same active ingredient, for spraying. This could be a dangerous concoction because chemical properties of pesticides can be changed when they are mixed, thereby increasing the detrimental effects (Salameh, Baldi, Brochard, *et al.* 2004). The above-mentioned farm practices are common during spraying operation in developing countries.

The level of safety awareness regarding pesticide use is low among farmers and this was cited as one of the major causes of pesticide toxicity in agrarian population in Ethiopia (Ejigu and Mekonnen 2005). Farmers are occupationally exposed to several potentially harmful environmental agents. Such agents include pesticides, fertilizers, fuels, engine exhaust, organic and inorganic dusts, solvents, ultraviolet light, and zoonotic bacteria and viruses. The factors contributing to the onset of pesticide-related cases are insufficient protective measures, followed by carelessness and inadequate information (Ejigu and Mekonnen 2005). Poisoning with organophosphates was the most common cause of intoxication in Turkey. The routes of intoxication are gastrointestinal and dermal; however, intravenous intoxication may occur even with small amounts of the poison (Guven, Unluhizarci, Goktas, *et al.* 1997).

The Food and Agriculture Organization and WHO recommend that Ia (extremely hazardous) and Ib (highly hazardous) pesticides should not be used in developing countries (PAN UK 2001). They also suggest that class II (moderately hazardous) pesticides be avoided. However, the practice of spraying these "powerful" pesticides continues despite the efforts (Table 2). By adopting aggressive marketing strategies, large chemical industries strengthen the myth that more potent pesticides are necessary to prevent crop loss, thus resulting in widespread usage of pesticides in the agriculture sector (Nigg, Beir, Carter, *et al.* 1990). Organophosphate pesticides are easily absorbed by respiratory and gastrointestinal mucosa, and lipophilic organophosphate compounds are readily absorbed through the skin (Karalliedde 1999).

Organophosphate pesticides commonly used in agriculture include malathion, parathion, dimethoate, chloropyrophos, monocrotophos, and quinalphos. The observed mortality rate varies between 4% and 30% following poisoning by organophosphate pesticides (Yamashita, Tanaka, and Ando 1997). Organophosphorus compounds are commonly used as insecticides, pesticides, and fungicides (Malik, Mubarik, and Romshoo 1998). Exposure to pesticides such as phosphamidon (55%), malathion (12.2%), dichlorvos (8.5%), Tic-20 (10.4%), and unknown (13.4%) were observed due to organophosphorus poisoning. The means of exposure to pesticides include ingestion (85.4%), inhalation (4.3%), and tropical application (10.4%). Seventy-five per cent patients use pesticides for suicide attempts, and 26% pesticide exposure may be accidental (Malik, Mubarik, and Romshoo 1998). The indiscriminate and increasing use of organophosphates as agricultural and household insecticides without educating the public on their storage and safe use potentially increases the risk of food poisoning outbreaks (Chaudhary, Lall, Mishra, *et al.* 1998).

Few reports from developing countries on the effects of pesticides on the general health of farming population are available from sources such as PubMed and ScienceDirect. Reports on which pesticide causes the most health problems are not available as spraying of mixed pesticides such as organophosphates, organochlorines, and carbamates is practised in most developing countries. In this chapter, the available literature on the adverse health effects of pesticides to agrarian population of developing countries and health surveys conducted by the Indian Institute of Toxicology Research (IITR, formerly Industrial Toxicology Research Centre) during 1990–2007 are discussed in detail.

Table 2 Pesticides commonly used in developing countries and their categorization by WHO hazard class

Pesticide (common name, WHO toxicity classification)	Chemical family
Ia Extremely hazardous	
Phorate	Organophosphate
Ib Highly hazardousa	
Monocrotophos	Organophosphate
Profenofos and Cypermethrin	Combination of an organophosphate and asynthetic pyrethroid
Carbofuran	Carbamate
II Moderately hazardous	
Dimethoate	Organophosphate
Quinalphos	Organophosphate
Endosulfan	Organochlorine
Carbaryl	Carbamate
Chlorpyrifos	Organophosphate
Cyhalothrin	Pyrethroid
Fenthion	Organophosphate
DDT	Organochlorine
III Slightly hazardous	
Malathion	Organophosphate
IV Unlikely to present acute hazard in normal use	
Carbendazim	Carbamate
Atrazine	Triazine

Source WHO (2004)

HEALTH IMPACTS

Neurological Effects

The brain and peripheral nervous system are directly affected by pesticides because they are the sites of action and deposition. All pesticide classes may affect brain and neural tissues even if they do not cause observable effects. The pesticides that most often affect the nervous system are the organophosphate and *N*-methyl carbamate insecticides; these pesticides are responsible for most acute poisonings. In addition to acute poisoning, organophosphates may produce sub-acute, delayed, and chronic neurological, neuro-behavioural, and psychiatric syndromes. Evidence for such chronic neurological and psychiatric effects of organophosphate compounds comes from case reports, clusters of neurological diseases, and studies of exposed workers and other populations (Moses 1995).

By alkyl phosphorylation of a serine hydroxyl group at the esteratic site, organophosphate compounds inactivate acetylcholinesterase. The phosphorylated enzyme becomes inactive and thus is unable to hydrolyse acetylcholine (Karalliedde 1999). This causes slowing down or stopping of nerve impulse transmission. Studies on neurological symptoms among Sri Lankan farmers occupationally exposed to acetylcholinesterase-inhibiting insecticides showed 24% acute pesticide poisoning cases. Significantly more inhibition of cholinesterase activity was found in farmers than that in controls (Smit, van-Wendel-de-Joode, Heederik, *et al.* 2003). A dose–effect relationship was found between the neuro-behavioural signs and symptoms and the use of multiple organochlorines (Kishi, Hirschhorn, Djajadisastra, *et al.* 1995).

A report (Peiris-John, Ruberu, Wickremasinghe, *et al.* 2002) on the effects of occupational exposure to organophosphate pesticides on nerve and neuromuscular function showed reduction in both sensory conduction velocity and motor conduction velocity in farmers between cultivation seasons in Sri Lanka. This study also showed evidence of sensory dysfunction after acute exposure and sensory and motor impairment after long-term, low-level exposure. Chronic neurological effects associated with pesticide exposure include slowing of nerve conduction velocity, slowing of motor/visual speed tests, slowing of reaction time, poorer performance of learning/memory tasks, impaired vibrotactile sensitivity, postural sway abnormality, increased beta activity in ECG, reduced amplitude in visually evoked potentials, and decreased muscle strength. Other neurological effects potentially associated with pesticide exposure are amyotrophic lateral sclerosis, eye disorders, Guillain–Barre syndrome, movement disorders, multiple system atrophy, psychiatric disorders, and reflex sympathetic dystrophy (Moses 1995). About 3% of the pesticide-exposed children showed signs of mental retardation and delayed milestones (Jamil, Das, Shaik, *et al.* 2007).

Le Couteur, McLean, Taylor, *et al.* (1999) suggested that pesticide exposure may be associated with increased risk of neurodegenerative diseases, particularly Parkinson's disease. In cross-sectional studies, an association was found between exposure to pesticides and Parkinson's disease, although no specific pesticides or their classes were identified (Le Couteur, McLean, Taylor, *et al.* 1999). Several studies have shown that the herbicide paraquat causes selective degeneration of neurons and induces parkinsonian symptoms (Liou, Tsai, Chen, *et al.* 1997). A number of case reports have described symptoms of Parkinson's disease in individuals exposed to organophosphates (Bhatt, Elias, and Mankodi 1999); herbicides, including glyphosate (Barbosa, Leiros, Costa, *et al.* 2001) and diquat (Sechi, Agnetti, Piredda, *et al.* 1992); and fungicides, including maneb (Meco, Bonifati, Vanacore, *et al.* 1994). In severe cases of poisoning by organophosphates such as chlorpyrifos, fenthion, malathion, and trichlorfon, muscle weakness, ataxia, and paralysis may occur after a period of apparent recovery. This condition is characterized by degeneration of axon and degeneration of myelin in the peripheral and central nervous systems and is known as organophosphate-induced delayed neuropathy. Such delayed neurotoxic effects are caused by binding (phosphorylation) of the enzyme neurotoxic esterase in nervous tissue, rather than inhibition of cholinesterase (Moses 1995).

Respiratory Effects

Reports of health problems among farm workers in Ethiopia show prevalence of respiratory symptoms of cough, phlegm, and wheezing (Ejigu and Mekonnen 2005). Insecticides, primarily those that inhibit cholinesterase, may cause respiratory symptoms among agricultural workers (Yemaneberhan, Bekele, Venn, *et al.* 1997; Ohayo-Mitoko, Kromhout, Simwa, *et al.* 2000). A study reported the prevalence of mild, moderate, and severe airway obstruction among pesticide sprayers (Kesavachandran, Singh, Mathur, *et al.* 2006b).

Dermal Effects

In Japan, allergic contact dermatitis was diagnosed in 274 cases (33.6%) during 1968–70 among all 815 pesticide-related diseases (including poisoning) registered during the period (Matsushita, Nomura, and Wakatsuki 1980). Among 122 fruit farmers in Taiwan who sprayed pesticides regularly, contact allergy to pesticides was observed in 40% and clinical symptoms of contact dermatitis were found in 30% (Guo, Wang, Lee, *et al.* 1996). In Spain, the prevalence of contact allergy to mercury and carbamates, which are compounds of many pesticides, was three times as high among farmers as it was among the control group (Garcia-Perez, Garcia-Bravo, and Beneit 1984). Among 104 Polish farmers treated for eczema, contact allergy to pesticides was found in two persons (Spiewak 2001). Similarly, among 263 hop growers in eastern Poland, 66 persons (25.1%) were found to have contact allergy to pesticides (Spiewak 2001). Pesticide-related contact dermatitis may induce skin irritation (Li 1986; Lisi, Caraffini, and Assalve 1987).

The International Agency for Research on Cancer has reported that professional pesticide sprayers are at increased risk of developing skin and lip cancer (IARC 1991). Strong carcinogenic properties were attributed to arsenical pesticides (Axelson 1987). Arsenic is a carcinogenic metal exhibiting clear predilection for inducing skin lesions (Lansdown 1995). In a study conducted in Costa Rica, high incidence of skin cancers (lip cancer, melanoma, non-melanocytic skin, and penile cancer) was found to occur in coffee growing regions, where paraquat and lead arsenate were extensively used (Wesseling, Antich, Hogstedt, *et al.* 1999).

Reproductive Abnormalities

Data collected on reproductive toxicity from couples engaged in spraying of insecticides such as organochlorine, organophosphate, and carbamate in cotton fields showed abnormal reproductive performance (Rupa, Reddy, and Reddy 1991). A study showed (Saiyed, Dewan, Bhatnagar, *et al.* 2003) that endosulfan exposure in male children delays sexual maturity and interferes with sex hormone synthesis.

Cancer

Risk of stomach cancer is found to be higher in farmers compared to the general population, with magnitude ranging from 1.05 to 1.12 (Acquavella, Olsen, Cole, *et al.* 1998; Blair, Zahm, Pearce, *et al.* 1992; Meyer, Chrisman, Moreira, *et al.* 2003). Several studies have indicated that farmers or agricultural workers may have an elevated risk of brain cancer (Blair, Zahm, Pearce, *et al.* 1992; Wingren and Axelson 1992; Brownson, Reif, Chang, *et al.* 1990; Reif, Pearce, and Fraser 1989; Musicco, Filippini, and Bordo 1982). Associations between environmental agents and brain cancer have not been fully evaluated; a number of pesticides have demonstrated carcinogenic potential in animal bioassays (Blair and Zahm 1993; IARC 1987; Hoover and Blair 1991).

General Health Problems

Liver function tests showed elevated values of alkaline phosphatase, glutamate pyruvate transaminase, and glutamate oxaloacetate transaminase in sprayers (Ejigu and Mekonnen 2005). The electrocardiography (ECG) examination of workers spraying methomyl, a carbamate insecticide, showed significant changes, indicating the cardiotoxic effect of methomyl

(Saiyed, Dewan, Bhatnagar, *et al.* 2003). Jamil, Das, Shaik, *et al.* (2007) conducted epidemiological studies in 200 pesticide-exposed agricultural workers and an equal number of age- and sex-matched controls. They found that 3% of the subjects showed decreased red blood cells and haemoglobin and increased white blood cells, as well as a large number of immature cells. Clinical manifestation observed among pesticide sprayers in an earlier study (Nagami, Nishigaki, Matsushima, *et al.* 2005) in Japan showed acute dermatitis (24%), chemical burns (15%), and eye disorders (6%).

Figures 1–3 show pesticide handling and spraying practices by farmers in India. Such common pesticide handling activities may provide evidence to readers of the risks routinely faced by farmers in developing countries.

Figure 1 Knapsack spraying in vegetable field

Figure 2 Tractor with pesticide-filled tanker

Figure 3 Pesticide spraying in mango plantation

HEALTH SURVEYS CONDUCTED BY IITR

A health study conducted by the IITR in Lucknow (ITRC 1990) on pesticide applicators (spraying a mixture of pesticides) working in Malihabad mango plantations showed overall morbidity rates of 42.8%. The chief morbidities were respiratory disorders (33.4%), musculoskeletal disorders (15%), and those pertaining to the central nervous system (6%), predominantly peripheral neuropathy. Singh, Jyoti, Reddy, *et al.* (2007) and Kesavachandran, Rastogi, and Mathur (2006a) carried out studies to determine health problems related to pesticide exposure among sprayers in mango orchards and observed 32.4% respiratory morbidity, 23.5% skin problems, 17.6% gastrointestinal effects, and 8.8% ocular problems, as well as inhibition of acetylcholinesterase and butyrylcholinesterase activities. They also found higher malondialdehyde levels among sprayers compared to controls.

Srivastava, Gupta, Bihari, *et al.* (1995) examined serum levels of thyroxin and thyroid stimulating hormone with respect to blood levels of organochlorine insecticides and observed depleted thyroxine levels in 24.3% subjects. Lung function abnormalities, especially prevalence of mild and moderate types of bronchial obstruction among sprayers, were observed in the earlier report (Rastogi, Gupta, Husain, *et al.* 1989). Higher nervous system functions, such as memory, learning, and vigilance, were also found to be affected in subjects exposed to quinalphos (Srivastava, Gupta, Bihari, *et al.* 2000). A significantly higher risk for sickness related to various systems—cardiovascular, genitourinary, respiratory, nervous, and dermal—was observed among retail pesticide shopkeepers compared to controls. These findings provide a prima facie evidence of clinical manifestations because of multiple exposures to pesticides and poor safety culture at workplace (Kesavachandran, Pathak, Fareed, *et al.* 2009a).

A recent study reported the risk of cholinesterase inhibition and health to be higher in tractor mounted pesticide sprayers than in knapsack sprayers and in both groups compared to controls. Occupational exposure among pesticide sprayers in North India needs better control, perhaps through redesigning spraying equipment (Pathak, Fareed, Bihari, *et al.* 2011b). A correlation is found between observed haematological abnormalities in sprayers and pesticide exposure, as evidenced by lower cholinesterase activity and the presence of organochlorine pesticides in blood (Fareed, Pathak, Bihari, *et al.* 2010). In another study, it was found that ergonomic factors, prolonged exposure to mixture of pesticides, and drop in cholinesterase activity may result in nerve conduction dysfunction (Pathak, Fareed, Bihari, *et al.* 2010a). A detailed evaluation of health in the population exposed to pesticides can be found in the reference Kesavachandran, Fareed, Pathak, *et al.* (2009b).

CONCLUSIONS

Pesticides commonly used in agriculture in developing countries are organochlorines and organophosphates. A considerable number of agrarian populations in developing countries continue to be exposed to pesticides. Owing to contact with pesticides in agricultural activities, farmers suffer from various health problems, such as neurological abnormalities, respiratory ailments, reproductive disorders, and dermal diseases. The literature from developing countries on the adverse health effects is limited and a few research groups, including IITR, in India have taken up initiatives to monitor the health problems in the pesticide-exposed agrarian community. The unique challenges in the farming sector of developing countries are illiteracy and poor knowledge of farmers about the toxic effects of pesticides. A sound national policy with appropriate guidelines for pest control is, therefore, essential in developing countries to reduce the adverse health consequences among the agrarian population.

REFERENCES

Acquavella, J., G. Olsen, P. Cole, B. Ireland, J. Kaneene, S. Schuman, and L. Holden. 1998. Cancer among farmers: a meta-analysis. *Annals of Epidemiology* 8: 64–74

Axelson, O. 1987. Pesticides and cancer risk in agriculture. *Medical Oncology and Tumor Pharmacotherapy* 4: 207–217

Barbosa, E. R., D. Leiros, M. D. Costa, L. A. Bacheschi, M. Scaff, and C. C. Leite. 2001. Parkinsonism after glycine-derivate exposure. *Movement Disorders* 16: 565–568

Bhalli, J. A., Q. M. Khan, M. A. Haq, A. M. Khalid, and A. Nasim. 2006. Cytogenetic analysis of Pakistani individuals occupationally exposed to pesticides in a pesticide production industry. *Mutagenesis* 21: 143–148

Bhatt, M. H., M. A. Elias, and A. K. Mankodi. 1999. Acute and reversible parkinsonism due to organophosphate pesticide intoxication: five cases. *Neurology* 52: 1467–1471

Blair, A. and S. Zahm. 1993. Patterns of pesticide use among farmers: implications for epidemiologic research. *Epidemiology* 4: 55–62

Blair, A., H. S. Zahm, N. Pearce, E. F. Heineman, and J. F. Fraumeni. 1992. Clues to cancer etiology from studies of farmers. *Scandinavian Journal of Work, Environment and Health* 18: 209–215

Brownson, R., J. Reif, J. Chang, and J. Davis. 1990. An analysis of occupational risks for brain cancer. *American Journal of Public Health* 80: 169–172

Bryant, D. H. 1985. Asthma due to insecticide sensitivity. Australian and New Zealand Journal of Medicine 15: 66–68

Cavaliere, M. J., E. E. Calore, N. M. Perez, and F. R. Puga. 1996. Organophosphate-induced myotoxicity. Revista de Saude Publica 30: 267–272

Chaudhary, R., S. B. Lall, B. Mishra, and B. Dhawan. 1998. A foodborne outbreak of organophosphate poisoning. *British Medical Journal* 317: 268–269

Chiang, L. W., Y. C. Chang, and D. J. Fang. 1998. The clinical significance of hyperamylasaemia in organophosphate poisoning. *Clinical Toxicology* 36: 673–681

Clarke, E. E. K., L. S. Levy, A. Spurgeon, and I. A. Calvert. 1997. The problems associated with pesticide use by irrigation workers in Ghana. *Occupational Medicines* 47: 301–308

Dasgupta, S., C. Meisner, D. Wheeler, K. Xuyen, and N. T. Lam. 2007. Pesticide poisoning of farm workers: implications of blood test results from Vietnam. *International Journal of Hygiene and Environmental Health* 210: 121–132

Dewan, A. and H. N. Saiyed. 1998. Acute poisonings due to agricultural pesticides reported to the NIOH Poison Information Centre. In *Proceedings of the WHO Workshop on Occupational Health Problems in Agriculture Sector*, edited by J. R. Parikh, V. N. Gokani, P. B. Doctor, D. N. Gandhi, and H. N. Saiyed. Ahmedabad: National Institute of Occupational Health

Eddleston, M. 2000. Patterns and problems of deliberate self-poisoning in the developing world. *Quarterly Journal of Medicine* 93: 715–731

Eddleston, M. and M. R. Phillips. 2004. Self-poisoning with pesticides. *British Medical Journal* 328: 42–44

Ejigu, D. and Y. Mekonnen. 2005. Pesticide use on agricultural fields and health problems in various activities. *East African Medical Journal* 82: 427–432

Ergonen, A. T., S. Salacin, and M. H. Ozdemir. 2005. Pesticide use among greenhouse workers in Turkey. *Journal of Clinical Forensic Medicine* 12: 205–208

Fareed, M., M. K. Pathak, V. Bihari, M. K. R. Mudium, D. K. Patel, N. Mathur, M. Kuddus, and C. N. Kesavachandran. 2010. Haematological and biochemical alterations in sprayers occupationally exposed to mixture of pesticides at a mango plantation in Lucknow, India. *Toxicological and Environmental Chemistry* 92: 1919–1928

Garcia-Perez, A., B. Garcia-Bravo, and J. V. Beneit. 1984. Standard patch tests in agricultural workers. *Contact Dermatitis* 10: 151–153

Government of India. 2001. *Tenth Five-year Plan: 2002–2007.* New Delhi: Planning Commission of India. Details available at <http://planningcommission.nic.in/plans/planrel/fiveyr/welcome.html>

Gunnell, D. and M. Eddleston. 2003. Suicide by intentional ingestion of pesticides: a continuing tragedy in developing countries. *Journal of International Epidemiology* 32: 902–909

Guo, Y. L., B. J. Wang, C. C. Lee, and J. D. Wang. 1996. Prevalence of dermatoses and skin sensitisation associated with use of pesticides in fruit farmers of southern Taiwan. *Occupational and Environmental Medicine* 53: 427–434

Gupta, P. K. 2004. Pesticide exposure: Indian scene. *Toxicology* 198: 83–90

Guven, M., K. Unluhizarci, Z. Goktas, and S. Kurtoglu. 1997. Intravenous organophosphate injection: an unusual way of poisoning. *Human and Experimental Toxicology* 16: 279–280

Hoover, R. and A. Blair. 1991. Pesticides and cancer. In *Cancer Prevention,* edited by V. Devita, S. Hellman, and S. Rosenberg. Philadelphia: Lippincott

IARC (International Agency for Research on Cancer). 1987. *IARC Monographs on the Evaluation of Carcinogenic Risks to Humans. Overall Evaluations of Carcinogenicity: an updating of IARC monographs.* France: IARC

IARC (International Agency for Research on Cancer). 1991. Occupational exposures in spraying and application of insecticides. *IARC Monographs on the Evaluation of the Carcinogenic Risk of Chemicals to Humans* 53: 45–92

Indra, D. P., R. Bellamy, and P. Shyamsundar. 2007. Facing hazards at work: agricultural workers and pesticide exposure in Kuttanad, Kerala. *South Asian Network for Development and Environmental Economics* 19: 1–4

ITRC (Industrial Toxicology Research Centre). 1990. *Epidemiological Health Survey of Pesticide Sprayers.* Lucknow: ITRC

Jamil, K., G. P. Das, A. P. Shaik, S. S. Dharmi, and M. Sudha. 2007. Epidemiological studies of pesticide-exposed individuals and their clinical implications. *Current Science* 92: 340–345

Jeyaratnam, J. 1985. Health problems of pesticide usage in the Third World. *British Journal of Industrial Medicine* 42: 505–506

Jeyaratnam, J. 1990. Pesticide poisoning: as a major global health problem. *World Health Statistics Quarterly* 43: 139–144

Jeyaratnam, J. and K. S. Chia. 1994. *Occupational Health in National Development.* Singapore: World Scientific

Karalliedde, L. 1999. Organophosphorus poisoning and anaesthesia. *Anaesthesia* 54: 1073–1088

Karunakaran, C. O. 1958. The Kerala food poisoning. *Journal of the Indian Medical Association* 31: 204–205

Kesavachandran, C., M. K. Pathak, M. Fareed, V. Bihari, N. Mathur, and A. K. Srivastava. 2009a. Health risks of employees working in pesticide retail shops: an exploratory study. *Indian Journal of Occupational and Environmental Medicine* 13: 121–126

Kesavachandran, C., M. Fareed, M. K. Pathak, V. Bihari, N. Mathur, and A. K. Srivastava. 2009b. Adverse health effects of pesticides in agrarian populations of developing countries. *Reviews of Environmental Contamination and Toxicology* 200: 33–48

Kesavachandran, C., S. K. Rastogi, and N. Mathur. 2006a. Health status among pesticide applicators at a mango plantation in India. *Journal of Pesticide Safety Education* 8: 1–9

Kesavachandran, C., V. K. Singh, N. Mathur, S. K. Rastogi, M. K. Siddiqui, M. M. Reddy, R. S. Bharti, and M. K. Asif. 2006b. Possible mechanism of pesticide toxicity related oxidative stress leading to airway narrowing. *Redox Report* 11: 159–162

Kishi, M., N. Hirschhorn, M. Djajadisastra, L. N. Satterlee, S. Strowman, and R. Dilts. 1995. Relationship of pesticide spraying to signs and symptoms in Indonesian farmers. *Scandinavian Journal of Work, Environment and Health* 21: 124–133

Konstantinova, M. K., C. Charlier, M. Craen, M. Du Caju, C. Heinrichs, C. D. Beaufort, G. Plomteux, and J. P. Bourguignon. 2001. Sexual precocity after migration from developing countries to Belgium: evidence to previous exposure to organochlorine pesticides. *Human Reproduction* 16: 1020–1026

Lansdown, A. B. 1995. Physiological and toxicological changes in the skin resulting from the action and interaction of metal ions. *Critical Reviews in Toxicology* 25: 397–462

Le Couteur, D. G., A. J. McLean, M. C. Taylor, B. L. Woodham, and P. G. Board. 1999. Pesticides and Parkinson's disease. *Biomedicine and Pharmacotherapy* 53: 122–130

Li, W. M. 1986. The role of pesticides in skin disease. *International Journal of Dermatology* 25: 295–297

Liou, H. H., M. C. Tsai, C. J. Chen, J. S. Jeng, Y. C. Chang, and S. Y. Chen. 1997. Environmental risk factors and Parkinson's disease: a case-control study in Taiwan. *Neurology* 48: 1583–1588

Lisi, P., S. Caraffini, and D. Assalve. 1987. Irritation and sensitization potential of pesticides. *Contact Dermatitis* 17: 212–218

London, L., A. J. Flisher, C. Wesseling, D. Mergler, and H. Kromhout. 2005. Suicide and exposure to organophosphate insecticides: cause or effect? *American Journal of Industrial Medicine* 47: 308–321

Malik, G. M., M. Mubarik, and G. J. Romshoo. 1998. Organophosphorus poisoning in the Kashmir Valley 1994–97. *New England Journal of Medicine* 338: 1078

Maria, C. P. R., X. P. Dario, and D. C. Eloisa. 2006. Acute poisoning with pesticides in the state of Mato Grosso do Sul, Brazil. *Science of the Total Environment* 357: 88–95

Matsushita, T., S. Nomura, and T. Wakatsuki. 1980. Epidemiology of contact dermatitis from pesticides in Japan. *Contact Dermatitis* 6: 255–259

McConnell, R., F. Pacheco, K. Wahlberg, W. Klein, O. Malespin, R. Magnotti, M. Akerblom, and D. Murray. 1999. Sub-clinical health effects of environmental pesticide contamination in a developing country: cholinesterase depression in children. *Environmental Research* 81: 87–91

McCormack, A. L., M. Thiruchelvam, A. B. Manning-Bog, C. Thiffault, J. W. Langston, D. A. Cory-Slechta, and D. A. Di Monte. 2002. Environmental risk factors and Parkinson's disease: selective degeneration of nigral dopaminergic neurons caused by the herbicide paraquat. Neurobiology of Disease 10: 119–127

Meco, G., V. Bonifati, N. Vanacore, and E. Fabrizio. 1994. Parkinsonism after chronic exposure to the fungicide maneb (manganese ethylene-bis-dithiocarbamate). *Scandinavian Journal of Work, Environment and Health* 20: 301–305

Meyer, A., J. Chrisman, J. C. Moreira, and S. Koifman. 2003. Cancer mortality among agricultural workers from Serrana Region, state of Rio de Janeiro, Brazil. *Environmental Research* 93: 264–271

Mishra, U. K., D. Nag, W. A. Khan, and P. K. Ray. 1988. A study of nerve conduction velocity, late responses and neuromuscular synapse functions in organophosphate workers in India. Archives of Toxicology 61: 496–500

Moses, M. 1995. *Designer Poisons: how to protect your health and home from toxic pesticides.* San Francisco, CA: Pesticide Education Centre

Musicco, M., G. Filippini, and E. Bordo. 1982. Gliomas and occupational exposure to carcinogens: case-control study. *American Journal of Epidemiology* 116: 782–790

Nag, D., C. C. Singh, and S. Senon. 1977. Epilepsy endemic due to benzene hexachloride. *Tropical and Geographical Medicine* 29: 229–232

Nagami, H., Y. Nishigaki, S. Matsushima, T. Matsushita, S. Asanuma, N. Yajima, M. Usuda, and M. Hirosawa. 2005. Hospital-based survey of pesticide poisoning in Japan, 1998–2002. *International Journal of Occupational and Environmental Health* 11: 180–184

Nigg, H. N., R. C. Beir, O. Carter, C. Chaisson, C. Franklin, T. Lavy, R. G. Lewis, P. Lombardo, J. F. McCarthy, K. T. Maddy, M. Moses, D. Norris, C. Peck, K. Skinner, and R. G. Tardiff. 1990. Exposure to pesticides. In *The Effects of Pesticides on Human Health, Advances in Modern Environmental Toxicology*, edited by S. R. Baker and C. F. Wilkinson. New York: Princeton Scientific

O'Malley, M. 1997. Clinical evaluation of pesticide exposure and poisonings. *Lancet* 349: 1161–1166

Ohayo-Mitoko, G., H. Kromhout, J. Simwa, J. Boleij, and D. Heederik. 2000. Self-reported symptoms and inhibition of acetylcholinesterase activity among Kenyan agricultural workers. *Occupational and Environmental Medicine* 57: 195–200

PAN UK (Pesticide Action UK). 2001. The lists of lists. Briefing paper 3. Details available at <www.pan-uk.org/briefing/Listof1.pdf>

Pathak, M. K., M. Fareed, V. Bihari, M. K. R. Mudium, D. K. Patel, N. Mathur, M. Kuddus, and C. Kesavachandran. 2011a. Nerve conduction studies in sprayers occupationally exposed to mixture of pesticides in a mango plantation in Lucknow, North India. *Toxicological and Environmental Chemistry* 93: 188–196

Pathak, M. K., M. Fareed, V. Bihari, N. Mathur, A. K. Srivastava, M. Kuddus, and K. C. Nair. 2011b. Cholinesterase levels and morbidity in pesticide sprayers in North India. *Occupational Medicine* 61: 512–514

Patricia, M., T. Partanen, C. Wesseling, V. Bravo, C. Ruepert, and I. Burstyn. 2005. Assessment of pesticide exposure in the agricultural population of Costa Rica. *Annals of Occupational Hygiene* 49: 375–384

Peiris-John, R. J., D. K. Ruberu, A. R. Wickremasinghe, L. A. Smit, and W. van der Hoek. 2002. Effects of occupational exposure to organophosphate pesticides on nerve and neuromuscular function. *Journal of Occupational and Environmental Medicine* 44: 352–357

Pesticide health effects on humans. Pesticide Safety Education Pgm. Cornell University, NY. Details available at <http://pmep.cce.cornell.edu/facts-slides-self/facts/gen-posaf-health.html>

Pinieri, E., J. E. Krige, P. C. Bornman, and D. M. Linton. 1997. Severe necrotizing pancreatitis caused by organophosphate poisoning. Journal of Clinical Gastroenterology 25: 463–465

Rao, S. C. H., V. Venkateswarlu, T. Surender, M. Eddleston, and N. A. Buckley. 2005. Pesticide poisoning in South Asia: opportunities for prevention and improved medical management. *Tropical Medicine and International Health* 10: 581–588

Rastogi, S. K., B. N. Gupta, T. Husain, N. Mathur, and N. Garg. 1989. Study on respiratory impairment among pesticide sprayers in mango plantations. *American Journal of Industrial Medicine* 16: 529–538

Reif, J., N. Pearce, and J. Fraser. 1989. Occupational risks for brain cancer: a New Zealand cancer registry-based study. *Journal of Occupational Medicine* 31: 863–867

Rupa, D. S., P. P. Reddy, and O. S. Reddy. 1991. Reproductive performance in population exposed to pesticides in cotton fields in India. *Environmental Research* 55: 23–126

Saiyed, H., A. Dewan, V. Bhatnagar, U. Shenoy, R. Shenoy, H. Rajmohan, K. Patel, R. Kashyap, P. Kulkarni, B. Rajan, and B. Lakkad. 2003. Effect of endosulfan on male reproductive development. *Environmental Health Perspectives* 111: 1958–1962

Salameh, P., I. Baldi, P. Brochard, and B. A. Saleh. 2004. Pesticides in Lebanon: a knowledge, attitude and practice survey. *Environmental Research* 94: 1–6

Sechi, G., V. Agnetti, M. Piredda, M. Canu, F. Deserra, and H. A. Omar. 1992. Acute and persistent parkinsonism after use of diquat. *Neurology* 42: 261–263

Senanayake, N. 1998. Organophosphours insecticide poisoning. *Ceylon Medical Journal* 305: 1502–1503

Senanayake, N. and L. Karalliedde. 1986. Acute poisoning in Sri Lanka: an overview. Ceylon Medical Journal 31: 61–71

Sethuraman, V. A. 1977. Case of BHC poisoning in a heifer calf. *Indian Veterinary Journal* 56: 486–487

Singh, V. K., L. Jyoti, M. M. Reddy, C. Kesavachandran, S. K. Rastogi, and M. K. J. Siddiqui. 2007. Biomonitoring of organochlorines, glutathione, lipid peroxidation and cholinesterase activity among pesticide sprayers in mango orchards. *Clinica Chimica Acta* 377: 268–272

Siwach, S. B. and A. Gupta. 1995. The profile of acute poisoning in Haryana-Rohtak study. Indian Journal of Physics 43: 756–759

Smit, L. A., B. N. van-Wendel-de-Joode, D. Heederik, R. J. Peiris-John, W. van der Hoek. 2003. Neurological symptoms among Sri Lankan farmers occupationally exposed to acetylcholinesterase-inhibiting insecticides. *American Journal of Industrial Medicine* 44: 254–264

Spiewak, R. 2001. Pesticides as a cause of occupational skin diseases in farmers. *Annals of Agricultural and Environmental Medicine* 8: 1–5

Srivastava, A. K., B. N. Gupta, V. Bihari, N. Mathur, B. S. Pangtey, R. S. Bharti, and M. M. Godbole. 1995. Organochlorine pesticide exposure and thyroid function: a study in human subjects. *Journal of Environmental Pathology, Toxicology and Oncology* 14: 107–110

Srivastava, A. K., B. N. Gupta, V. Bihari, N. Mathur, L. P. Srivastava, B. S. Pangtey, R. S. Bharti, and P. Kumar. 2000. Clinical, biochemical and neurobehavioral studies of workers engaged in the manufacture of quinalphos. *Food and Chemical Toxicology* 38: 65–69

Wesseling, C., D. Antich, C. Hogstedt, A. C. Rodriguez, and A. Ahlbom. 1999. Geographical differences of cancer incidence in Costa Rica in relation to environmental and occupational pesticide exposure. *Journal of International Epidemiology* 28: 365–374

WHO (World Health Organization). 1990. Public Health Impact of Pesticides Used in Agriculture. Geneva: WHO

WHO (World Health Organization). 2004. The Recommended Classification of Pesticides by Hazard and Guidelines to Classification. Geneva: World Health Organization. Details available at <www.who.int/ipcs/publications/pesticides_hazard/en/>

Wingren, G. and O. Axelson. 1992. Cluster of brain cancers spuriously suggesting occupational risk among glassworkers. *Scandinavian Journal of Work, Environment and Health* 18: 85–89

Yamashita, M., J. Tanaka, and Y. Ando. 1997. Human mortality in organophosphate poisonings. *Veterinary and Human Toxicology* 39: 84–85

Yassi, Y. A., T. Kjellstrom, T. K. Kok, and T. L. Gudotli. 2001. *Basic Environmental Health*, World Health Organization. London: Oxford University Press

Yemaneberhan, H., Z. Bekele, A. Venn, S. Lewis, E. Parry, and J. Britton. 1997. Prevalence of wheeze and asthma and relation to atopy in urban and rural Ethiopia. *Lancet* 350: 85–90

CHAPTER 7

CURRENT LAWS AND LEGISLATION IN INDIA

In India, pesticides are regulated by different acts and rules and various government agencies are involved in the regulation of pesticides. These laws and acts regulate the manufacture and use of pesticides in the country.

THE PLANTATIONS LABOUR (AMENDMENT) ACT, 2010 (CHAPTER IVA, PROVISIONS AS TO SAFETY)

(1) In every plantation, effective arrangements shall be made by the employer to provide for the safety of workers in connection with the use, handling, storage and transport of insecticides, chemicals and toxic substances.

(2) The State Government may make rules for prohibiting or restricting employment of women or adolescents in using or handling hazardous chemicals.

(3) The employer shall appoint persons possessing the prescribed qualifications to supervise the use, handling, storage and transportation of insecticides, chemicals and toxic substances in his plantation.

(4) Every employer shall ensure that every worker in plantation employed for handling, mixing, blending and applying insecticides, chemicals and toxic substances is trained about the hazards involved in different operations in which he is engaged, the various safety measures and safe work practices to be adopted in emergencies arising from spillage of such insecticides, chemicals and toxic substances and such other matters as may be prescribed by the State Government.

(5) Every worker who is exposed to insecticides, chemicals, and toxic substances shall be medically examined periodically, in such a manner as may be prescribed, by the State Government.

(6) Every employer shall maintain health record of every worker who is exposed to insecticides, chemicals, and toxic substances which are used, handled, stored, or transported in a plantation, and every such worker shall have access to such a record.

(7) Every employer shall provide:

(a) washing, bathing, and clock room facilities; and

(b) protective clothing and equipment, to every worker engaged in handling insecticides, chemicals, or toxic substances in such a manner as may be prescribed by the State Government.

(8) Every employer shall display in the plantation a list of permissible concentrations of insecticides, chemicals, and toxic substances in the breathing zone of the workers engaged in the handling and application of such insecticides, chemicals, and toxic substances.

(9) Every employer shall exhibit such precautionary notices as may be prescribed by the State Government indicating the hazards of insecticides, chemicals, and toxic substances.

Furthermore, the law authorizes the State Government to make rules regarding the training of workers for safety and health at work, preventive medical examination of workers, facilities, and personal protective equipment to be provided to workers engaged in handling insecticides, chemicals, and toxic substances (The Plantations Labour (Amendment) Act 2010).

The new legislation would regulate hours of work and provide for health, safety, and welfare of plantation workers. Once the new law comes into force, the plantation employers would be obligated by Workmen's Compensation Act (Press Trust of India 2010).

THE ENVIRONMENT PROTECTION ACT, 1986

The Environment Protection Act (EPA), 1986 under the Ministry of Environment and Forests covers several rules applying to insecticides, such as the "Manufacture, Import and Storage of Hazardous Chemicals Rules, 1989". The major objective of this regulation is to prevent accidents and disasters. Public Liability Insurance Act, 1992 under the Ministry of Environment and Forests also applies to pesticides. There are other rules under the EPA, such as the Hazardous Waste (Management and Handling)

Rules, 1989, Water (Prevention and Control of Pollution) Act, 1974, and Air (Prevention and Control of Pollution) Act, 1981, that also apply to pesticides (Hagedorn and Chennamaneni 2011).

THE INSECTICIDES ACT, 1968 AND RULES, 1971

According to Section 3(e) of Insecticides Act, 1968, the word "insecticides" means (i) any substance specified in the Schedule; or (ii) such other substances (including fungicides and weedicides) as the Central Government may, after consultation with the Board, by notification in the Official Gazette, include in the schedule from time to time; or (iii) any preparation containing any one or more of such substances. Thus, technically all insecticides (pesticides) in India are those substances that are listed in the "Schedule" of the Insecticides Act, 1968. The Schedule contains an exhaustive list of insecticides which are legally covered under the Act (India Juris 2011). This Act pertains to regulation of import, manufacture, transport, distribution, sale, and use of insecticides with a view to prevent risk to human beings or animals and matters connected therewith (Mistry 2011). The salient features of Insecticide Rules, 1971 are discussed in the following sections (Mistry 2011).

Packaging and Labelling (Rules 16–20)

The packaging container used should be an approved one. The container should contain a leaflet that provides information such as the common name of the insecticide, plant disease for which it should be applied, manner of application, health effects of poisoning, suitable and adequate safety measures and necessary first-aid treatment, antidote, decontamination or safe disposal of the used container, storage, handling precautions, and effects on skin, nose, eye, throat, and so on.

The following warning statements should prominently appear on the labels of different categories:

- For category I (extremely toxic) insecticides, the symbol of a skull and crossbones should appear on the label, and the word "poison" should be printed in red. The warning statements "keep out of the reach of children" and "if swallowed or if symptoms of poisoning occur, call physician immediately" should appear on the label.
- For category II (highly toxic) insecticides, the word "poison" in red and statement "keep out of the reach of children" should appear.
- For category III (moderately toxic), the word "danger" and statement "keep out of the reach of children" should be added.

- For category IV (slightly toxic), the word "caution" should be mentioned. Category classification of insecticides is given in Table 1.

Transport and Storage (Rules 35 and 36)

In the case of rail transport, packages containing insecticides should be packed according to the "Red Tariff" rules. During transportation or storage, care should be taken to ensure that insecticides transported do not come in contact with foodstuffs or animal feeds. Competent authorities notified by the state government should examine the possible contamination in the event of any mixing up of insecticides with foodstuffs or animal feeds due to any damage to packages during transport or storage. The transport agency or the storage owner should take measures for safe disposal if insecticides are found to have leaked out during transport and storage in order to prevent poisoning and pollution of soil, water, and so on. The packages containing the insecticides should be stored in separate rooms or almirahs under lock and key. Such rooms shall be well built, well lit, dry, and ventilated and should have sufficient space.

Protective Equipment and Other Facilities for Workers (Rules 37–44)

All persons engaged in handling, dealing, or otherwise coming in contact with insecticides during manufacture, formulation, or spraying shall be medically examined before their employment and then periodically once in a quarter by a qualified doctor. The blood cholinesterase level of persons handling organophosphorous or carbamate compounds should be monitored on a monthly basis. The blood residue estimation of persons exposed to organochlorine pesticides should be done on a yearly basis. Any person showing symptoms of poisoning shall be immediately examined and given proper treatment.

Before a doctor is called to examine a person, the person must be given first-aid treatment. Workers should be educated regarding effects of poisoning and the first-aid treatment to be given. Protective clothing should

Table 1 Categories of insecticides

Classification of insecticides	Oral route (acute toxicity), LD_{50} mg/kg of test animal	Dermal route (dermal toxicity), LD_{50} mg/kg of test animal	Colour of band on the label
Extremely toxic	1–50	1–200	Bright red
Highly toxic	51–500	201–2,000	Bright yellow
Moderately toxic	501–5,000	2,001–20,000	Bright blue
Slightly toxic	>5,000	>20,000	Bright Green

be washable (to remove toxic exposure) and should prevent insecticide penetration. A complete suit should include protective overgarment/ overalls/hood/hat, rubber gloves extending halfway up to forearm, dust-proof goggles, and boots. For preventing inhalation of toxic dusts, vapours, and gases, workers shall use chemical cartridge respirator, supplied air respirator, full- or half-face gas mask with canister, and demand flow-type respirator.

Workplace should have sufficient stocking of first-aid tools, antidotes, equipment, medicines, and so on. Workers should be provided proper training in safety precautions and use of safety equipment. Packages and surplus materials shall be safely washed and disposed to prevent pollution. Reuse of packages must be prevented. Packages should be broken and buried away from habitation (Mistry 2011).

Responsibilities of Central and State Governments

The central government is responsible for pesticide registration and state governments are responsible for enforcement of the regulations relating to manufacture, transport, sale, distribution, and use. The state governments issue manufacture and sale licenses. Both central and state governments are responsible for quality control, and laboratories are available at centre, state, and regional levels for this purpose (Agricultural Legislations 2011).

There are 49 State Pesticide Testing Laboratories (SPTLs), 2 Regional Pesticide Testing Laboratories (RPTLs), and the Central Insecticides Laboratory, established under Section 16 of the Insecticides Act, 1968 to test and analyse the quality of insecticides. About 50,000 samples are tested annually. SPTLs are located in 20 states and 1 union territory, with a total annual analysis capacity of 51,440 samples. Besides, two RPTLs have also been set up in Kanpur and Chandigarh for supplementing the resources of the states/union territories in the quality control testing of pesticides, particularly for those states/union territories that lack SPTL.

At the central level, Central Insecticides Laboratory has been set up under Section 16 of the Insecticides Act, 1968 to perform the statutory role of referral analysis. Therefore, both central and state governments implement the Act (Agricultural Legislations 2011).

Central Insecticides Laboratory was set up under Section 16 of the Insecticides Act, 1968 with major objectives of pre- and post-registration verification of properties, performance, and hazards of pesticides and the proposed use claimed by manufacturers. Central Insecticides Board (CIB) was constituted under Section 4 of the Insecticides Act to advise central and state governments on technical matters arising out of administration of this Act. These include matters pertaining to risk to human beings or animals involved in the use of insecticides and safety measures necessary to prevent such risk and matters relating to manufacture, sale, storage, transport, and distribution of insecticides. Registration Committee (RC) was constituted under Section 5 of Insecticides Act, 1968 to scrutinize formulae of pesticides, verify claims with respect to efficacy and safety to human beings and animals, specify doses, and take precautions against poisoning and these details should be indicated on the label and leaflets. Registration Committee, under Section 9 of Act, registers a pesticide after verifying its efficacy and safety to human beings, animals, and environment (Ministry of Agriculture 2011).

All insecticides (pesticides) have to necessarily undergo the registration process with the Central Insecticides Board and Registration Committee (CIB&RC) before they can be made available for use or sale (India Juris 2011). Following the grant of registration, a prospective manufacturer is required to obtain a license to manufacture a particular pesticide from the state government where the manufacturing unit is located. After making proper inspection and ensuring that the essential infrastructure is in place, a license is granted to manufacture a pesticide. After ensuring technical competence of the operator, the Plant Protection Adviser to the Government of India issues licenses for commercial pest control (Singh 2011).

The Registration Certificate mandates that a label be put on the packaging which clearly indicates the nature of the insecticide (agricultural or household use), composition, active ingredient, recommended dosage, target pest(s), caution sign, and safety instructions. A pesticide labelled for agriculture should not be used in a household (India Juris 2011). Figure 1 shows the registration protocol for pesticides (Mueller 2011).

Up to 25 September 2008, a total of 215 pesticides have been registered for use in India. There are 25 pesticides banned for manufacture, import, and use, 2 pesticides or pesticide formulations banned for use but their manufacture is allowed for export, 4 pesticide formulations banned for import, manufacture, and use, and 8 pesticides withdrawn (Tata Chemicals Limited 2009). Some are restricted use pesticides, which means they are used only for prescribed purposes and by authorized personnel by obtaining the appropriate Government license (India Juris 2011). In the schedule to Insecticides Act, about 815 molecules are included. About 185 maximum residue level (MRL) set and several MRLs have to be established in the coming years (Mueller 2011).

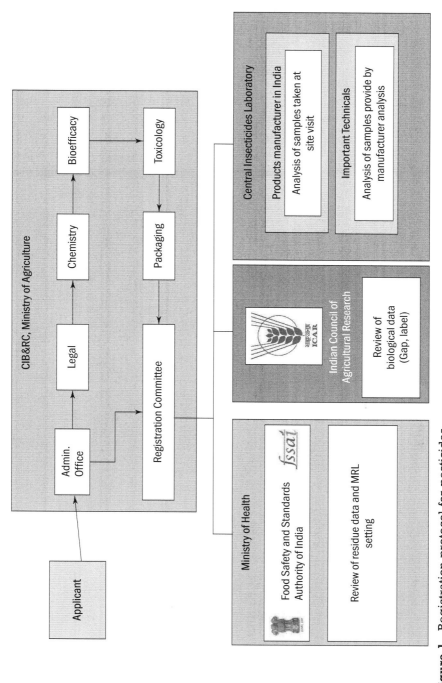

Figure 1 Registration protocol for pesticides

Pesticides Management Bill, 2008

On 24 April 2008, the Union Cabinet gave its approval in the parliament for introducing the Pesticides Management Bill, 2008 in order to replace the Insecticide Act, 1968. The bill aims at improving the quality of pesticides available to Indian farmers by introducing new, safe, and efficacious pesticides. The bill seeks effective regulation of manufacture, transport, distribution, import, export, sale, and use of pesticides to prevent the risk to human beings, animals, or environment and to de-license retail sale of household insecticides. Moreover, the bill promotes a detailed categorization of offences and punishments for greater deterrence to violators and timely disposal of time-barred pesticides in an environmentally safe manner (Tata Chemicals Limited 2009; India Juris 2011). The bill will regulate a wider range of chemicals, including weedicides and fungicides along with insecticides. Although there is a lot of optimism for the Pesticides Management Bill, 2008, the bill has not been able to make human and environmental health a priority (Pesticide Management Bill 2008).

However, there are various aspects of the Pesticide Management Bill, 2008 that require review and evaluation, are discussed next (Bhushan 2011). The bill should be renamed as Pesticide (Production and Distribution) Management Bill, 2008. The bill should focus on reducing the impact of chemical pesticides on natural resources, including land, water, and all living species; preventing use of pesticides for purposes other than crop protection; making an effective use of pesticides; reducing the amounts of crop residue remains; and curtailing malpractices in production, research, marketing, and distribution of pesticides. The major objective of this bill should be to control and regulate the process of production, marketing, sale, and distribution of pesticides. This bill should not be burdened with the mandate to control pests. The board needs to be more actively involved in regulation and management of production, sale, and distribution of pesticides. At present, its role is only advisory in nature. The board should have the power to review the decisions of the RC and to define its functions and composition. Member Secretary should be from the Ministry of Commerce and Industry or Ministry of Environment and not from the Directorate of Plant Protection. There should be a stipulated period for registrations and submission of information by the applicant. These registrations should be reviewed by an independent body with due consultations periodically.

In Chapter I, the bill does not mention that it replaces the existing act (1962). Moreover, the words "shall be in addition to" should probably be replaced by "as mentioned in the Environment Protection Act". Definitions should include "all living species" instead of just "animals" as pesticides affect all living species, including animals, birds, and bacteria. The section on definitions needs some revision, especially for definitions such as misbranding and so on. There should not be any deemed registered pesticide as scientific research reveals new evidences on previously registered pesticides. All registered and unregistered pesticides prior to this act should also undergo a similar registration process. According to Chapter III, Section 12(3) of the bill, applicants are required to furnish information "on all the known inimical effects of the pesticide". It should be changed to "on all the inimical effects of the pesticide" on the date of application and later on periodically as the knowledge of science unfolds. The bill should prescribe a minimum fee in terms of percentage of the turnover or economic value of production (in Chapter III, Section 12(4)). In Chapter III, Section 12(6) of the bill, it may be added that "any registration can be reviewed by the RC if fresh evidence through proper research data is provided by a body of persons, or individuals, in public interest". In Chapter III, Section 12(8), there is a provision for registration without providing proper data. This provision should be removed because hazardous substances without evidence cannot be permitted for sale and distribution as there will be scope for misuse. In Chapter III, Section 11(2), there should be incorporation of another function for the committee, that is, "to maintain a database of scientific and other research on the efficacy and inimical effects of various pesticides across the world". In Chapter IV, Grant of Licenses, the provision for a single licensing officer has drawbacks, as the officer might be overburdened and would not be in a position to arrive at suitable decisions. Instead a board should be constituted at the State Government level with functions similar to RC for licensing. In Chapter IV, Grant of Licenses 21(1), accreditation of private laboratories should be the function of the Central Pesticide Board (CPB). The accreditation process has to be established by the Board and not by the plant protection adviser. Chapter V should contain the following provision: "Any pesticide that is registered in India but is banned in any other country has to be examined with the available evidence and the license for its production and distribution has to be decided upon accordingly". The import of the pesticide should be restricted from the country in which it has been banned from production, distribution, or usage. In Chapter VII, offences and punishment, the penal provision should be raised from ₹25,000, which is far less to deter people from committing any offence. There should be different penalty rates at different levels. At the retail level, it should be about ₹50,000, and the amount has to increase with the increase in the geographical area. At the level of manufacturers, it should be about 10 per cent of the total turnover. The same approach should be adopted for imprisonment. In Chapter VII, Section 41 is very weak. Companies have to own the liability for the effects of the pesticides.

There are other additional issues about the bill that need consideration (Bhushan 2011). They include the following:

1. India is a party to several conventions by consent and participation. This bill should include commitment to such conventions. Central Pesticide Board should implement and enforce all the International Conventions in which the Government of India has become a party to.

2. Criminal liability should be imposed on manufacturers, distributors, and marketers as many times pesticides have been found to be misused.

3. Licensing should be made mandatory in the entire supply chain of pesticides from manufacturers to retailers.

4. There should be provision for checking pesticide prescriptions written by agricultural officers.

5. The bill should also contain clauses on emergency management, especially for hazardous situations, such as fire accidents in the pesticide factories, spillovers on land and water, and pesticide tanker collisions. It should also include local authorities such as panchayats, municipal bodies, fire services, and emergency medical services. There should also be provisions for prior information and emergency preparedness and response.

THE PREVENTION OF FOOD ADULTERATION ACT, 1954

The responsibility for production and use of insecticides in India lies with the Ministry of Health and Family Welfare. The legislation called Prevention of Food Adulteration (PFA) Act, 1954 was enacted for this purpose. Standards for various articles of food are prescribed under this act. This act also contains provisions for the storage, distribution, and sale of food items. The MRLs for different pesticides are prescribed under the PFA Act. The Central Committee for Food Standards (CCFS) constituted under Section 3 of the Prevention of Food Adulteration Act, 1954 advises central or state governments on all matters arising out of implementation of the Prevention of Food Adulteration Act, 1954 and the PFA Rules, 1955, including review and formulation of rules, regulations, and standards of food articles (Hagedorn and Chennamaneni 2011). In the Part XIV of the

Prevention of Food Adulteration Rules, 1955, Rule 65 (subrule 2) states the restriction on the use of insecticides directly on articles of food (The Prevention of Food Adulteration Acts and Rules 2011).

Tolerance levels have not been fixed under the PFA Act for many pesticides which have been approved for use in the country. Currently, 165 pesticides are approved for use, but tolerance levels for only 71 pesticides have been notified under Rule 65 of the PFA Rules. There are pesticides which were approved before 1971 and for which no data are available for assessing risk from the point of view of food safety and for fixing MRL. These pesticides are referred to as "deemed pesticides" and are not included under the PFA Act (Hagedorn and Chennamaneni 2011).

THE FACTORIES ACT, 1948

The Factories Act, 1948 under the Ministry of Labour deals with the safety aspects of pesticide production. This act contains 12 chapters which deal with health, safety, welfare, precautions undertaken in the case of hazardous processes, working hours, employment of women, young persons and children, leave, penalties, and so on (Hagedorn and Chennamaneni 2011; Planning Commission 2011). This act provides for health, safety, and well-being of all workers in factories; safety of life and limb of workers and provisions for hazardous process; ensuring welfare facilities; regulation of work hours for adults and adolescents; and employment of young persons and leave provisions (Factories Act in India 2011).

Effective enforcement of pesticide laws can protect public health and reduce environmental incidents involving pesticides among agricultural labourers, industrial workers, retailers, and stockists (Kesavachandran, Pathak, Fareed, *et al.* 2009a; Kesavachandran, Pathak, Fareed, *et al.* 2009b). The central government is empowered to legislate, register, and make acts for pesticides. The state governments are responsible for the enforcement of registration, legislation, and acts for pesticides.

The Ministry of Agriculture has the legislative powers to regulate pesticides. However, pesticides can cause harm to humans, animals, and the environment. Hence, the legislative power of the act should be with the Ministry of Health and the Ministry of Environment and Forest. Both the ministries should have the power to make decisions on registration, review, and ban of pesticides (Misra 2011).

As agriculture is a state subject, states should have the complete power to ban, restrict, or allow pesticide use. According to the Insecticides Act, 1968, the state could temporarily ban a pesticide for 60 days and

Act, 1968, the state could temporarily ban a pesticide for 60 days and then extend it to another 30 days, pending a study. In the Pesticide Management Bill, these limits have been extended to 180 days and 60 days, respectively (Misra 2011). To help resolve the pesticide issues related to health and environment, more powers need to be transferred from the central government to the state governments and there should be fast approval of legislations in a state assembly to adopt the pesticide management bill.

INSECTICIDE (AMENDEMENT) RULES, 2006

Published in the Gazette of India (extraordinary) Part II, Section 3, sub-section (i) vide G.S.R. 548(E), dated 12 September 2006

To exercise the powers conferred by Section 36 of the Insecticides Act, 1968 (46 of 1968), the Central Government, after consultation with the Central Insecticides Board, makes the following rules, further to amend the Insecticides Rules, 1971:

"1. These rules may be called the Insecticides (amendment) Rules, 2006. They shall come into force on the date of their publication in the official gazette.

2. In the Insecticide Rules, 1971 (hereinafter referred to as the said rules), in rule 2, for clauses (a) to (s), the following clauses shall be substituted:

 (a) "Act" means the Insecticides Act, 1968 (46 of 1968);

 (b) "Commercial Pest Control Operation" means any application or dispersion of insecticide including fumigants in household or public or private premises or land and includes pest control operations in the fields including aerial applications for commercial purposes but excludes private use;

 (c) "dealer" means a person carrying on the business of selling insecticides, whether wholesale or retail, and includes an agent of a dealer;

 (d) "expiry date" means the date that is mentioned on the container, label or wrapper of an insecticide against the column "date of expiry";

 (e) "Form" means a form set out in the First Schedule;

 (f) "Laboratory" means the Central Insecticides Laboratory;

(g) "manufacturer" means any person who manufactures insecticides and includes a formulator of the insecticides;

(h) "pests" means any insect, rodent, fungi, weed, and other forms of plant or animal life not useful to human beings;

(i) "Pest Control Operator" means any person who undertakes pest control operations and includes the person or the firm or the company or the organisation under whose control such a person is operating;

(j) "primary package" means the immediate package containing the insecticides;

(k) "principal" means the importer or manufacturer of insecticides, as the case may be;

(l) "registration" includes provisional registration;

(m) "rural area" means an area which falls outside the limits of any Municipal Corporation or Municipal Committee or a Notified Area Committee or a Cantonment;

(n) "Schedule" means a Schedule annexed to these rules;

(o) "secondary package" means a package which is neither a primary package nor transportation package;

(p) "section" means a section of the act;

(q) "testing facility" means an operational unit where the experimental studies are being carried out or have been carried out in relation to submission of data on product quality or on safety or on efficacy or on residues or on stability in storage of the insecticides for which an application for registration is made;

(r) "transportation package" means the outermost packaging used for transportation of insecticides.

3. After rule 10C of the said rules, the following rules shall be inserted, namely –

(a) "10D manufacturer and dealer" to display stock and price list of insecticides. Every manufacturer and every dealer who makes or offers to make a retail sale of any insecticide shall prominently display at his place of business,

(b) the amount of stock of different insecticides held by him on daily basis, and

(c) a list of prices or rates of such insecticides for the time being available in such place of business.

4. For rule 15 of the said rules, the following rule shall be substituted, namely–issuing of cash or credit memo and maintenance of register, books of accounts and records.

   (1) All sales of insecticides shall be made by a bill, cash memo, or credit memo in the form prescribed under any law.

   (2) All sales of insecticides made to a licensed manufacturer (formulator or packer), stockist, distributor, dealer, retailer, or to a bulk consumer shall be entered insecticide-wise, in a register in Form XII and a Statewise monthly return of all sales to actual consumers and shall be sent to the licensing officer, in Form XIV within 15 days from the close of the relevant month.

   (3) Every importer or manufacturer of insecticide shall maintain a stock register in Form XV for technical grade insecticides and in Form XVI for formulated insecticides.

   (4) Every manufacturer or importer shall maintain a book of accounts and register of such sale, manufacture, or import of technical Grade Insecticide and formulated Insecticide and submit the monthly statement or return to the licensing officer, in Form XV A and XVI A respectively within 15 days from the closing of the month.

   (5) Without prejudice to the provisions of sub-rules (a) and (d), the Central Government or the State Government or any other person authorized by the Central Government or the State Government, as the case may be, may, by notice in writing require any importer or manufacturer or any other person dealing in insecticides to furnish within the time specified in the notice, such registers, books of accounts, records, documents or other such information with respect to any insecticides or any batch thereof, including the particulars of all persons to whom it has been sold or distributed, as it may consider necessary."

In India, the Insecticide Act, 1968 and Insecticide Rules, 1971, administered by the Ministry of Agriculture, regulates the manufacture, import, sale, transport, distribution, and use of insecticides, with the intent of preventing risk to human beings or animals and other matters connected therewith. The CIB constituted under Section 4 of this Act advises central and state governments of India on technical matters, namely, specifies safety measures necessary to prevent risk to human beings or animals in manufacture, sale, storage, distribution, and use, assesses suitability for aerial application, sets specifications for shelf life, advises on residue tolerance limits and waiting periods, recommends inclusion of chemicals or substances in the schedule of insecticide, and pertains to other functions incidental to these matters. Effective implementation and enforcement of these laws (including Plantation Labour (Amendment) Bill, 2008) can prevent public health problems and reduce environmental incidents resulting from exposure to pesticides.

Improved safety may also be achieved by reducing the quantity of pesticides used in agriculture and by using personal protective equipment (PPE) or personal protective gadgets during spraying operations. This can be achieved either through restricting the concentration of formulations or by increasing the dilution of the spray stream by increasing recommended water spray volumes. There is also ample scope for reducing pesticide exposure to applicators through creation of improved pesticide toxicity awareness programmes. Such programmes are consistent with integrated pest management, which is an eco-friendly approach for pest management that includes cultural, mechanical, and biological methods as well as the need-based use of chemical pesticides, with preference given to the use of biopesticides and biocontrol agents. The use of integrated pest management has presumably reduced chemical pesticide consumption from 65,462 tons during 1994–95 to 47,020 tons during 2001–02 in India.

Pesticide users would benefit from the programmes initiated by State Department of Agriculture. The welfare fund board for agricultural labourers could initiate programmes for pesticide applicators, including health insurance protection measures. A programme focusing on improving awareness regarding pesticide toxicity and provision of a subsidy for protective equipment would contribute significantly towards health improvement among pesticide applicators. Efforts should be made to improve the safety of pesticide application, including provision of protective gear. Further research is required to develop indicators and methods for biological monitoring of occupational exposure to pesticides. For assessing the extent of exposure and absorption, pre-exposure value and reference value of relevant indicators are necessary. Providing health care and research facilities in the field of clinical neurosciences in rural areas will help share knowledge and facilities by both experts and people. With respect to pesticide usage, access to occupational health services is imperative. Agriculture should be declared as an industry and agricultural workers should be treated as industrial workers.

CONCLUSIONS

In India, 41 pesticides or formulations have been banned, including pesticide formulations banned for import, manufacture for export and pesticides withdrawn. Eighteen pesticides were refused for registration and 14 pesticides restricted for use in the country (http://cibrc.nic.in/list_pest_bann.htm). Specific national or international guidelines should be laid for pesticide usage. Before making a decision to ban a pesticide, the scientific understanding of the pesticide and related environmental issues should be considered in order to avoid controversies such as the one mentioned in Edwards (2004).

With increase in human population and crop production, there has been a dramatic rise in pesticide production across the world. The indiscriminate use of pesticides for crop protection against pest has resulted in a considerable human health hazard. The worldwide consumption pattern of pesticides is found to have undergone significant changes since the 1960s. Over the years, there has been a rapid increase in the use of herbicides in pesticide consumption and the consumption of insecticides, fungicides, and bactericides has declined. Pesticide production and use in developing countries such as India, China, and other Asian and African countries has increased exponentially to meet the food production requirement. The pesticide regulatory/legal measures are not strictly implemented in developing country like India. This has resulted in large-scale use of pesticides, especially in agricultural fields. The less awareness among agrarian population about the toxicity of pesticides to human health has resulted in adverse health consequences, especially in developing countries in Asia and Africa. Although the label on the pesticide bottle indicates the level of toxicity, it contains no specific details about the use of the pesticide and the disposal of cans. This has resulted in improper spraying methods and disposal of pesticide bottles or cans in the agricultural field, which further leachate to soil and lead to large-scale pollution of our ecosystem.

Unlike their predecessors, organochlorines, organophosphates and carbamates have a shorter environmental half-life and many of them act on a very broad spectrum. Presence of pesticides in the atmosphere is largely owing to emissions from pesticide industries, evaporation of pesticide residues from soils and water bodies, volatilization of pesticides sprayed, and so on. Generally, atmospheric pollution of pesticides is widespread. Organochlorine pesticides were detected even on Nanjiabawa peak in Tibet, which has an elevation of 4250 m (Shan 1997). The banned pesticide such, as DDT, which is used for malaria eradication and other national health programmes, has entered the different matrices of environment, resulting in pollution and entry in human system through food, water, and so on.

Pesticide exposure can lead to acute or chronic health effects. Acute (or short-term) effects generally occur immediately after heavy exposure to pesticides. A chronic effect develops slowly over a long period of time and may last for several years after exposure. The effects result from long-term, low-dose repeated exposure to a pesticide. The effects may also result from exposure to a high dosage over a short time. Chronic health effects lead to cancer, interfere with foetal and child growth, and disrupt reproductive, endocrine, immune, and central nervous systems (neurotoxic effects). Pesticides are suspected endocrine disruptors and linked to cancer development, abnormal sexual development, endometriosis, lowered male fertility, damage to thyroid and pituitary glands, lowered immunity, and behavioural problems.

Over the past several years, pesticide pollution of soil, water and air and pesticide-related deaths in developing countries have been a serious health concern. The Blacksmith Institute, New York has identified pesticide as one of the most severe human health toxicants. Most health studies on population exposed to pesticides have focused on farmers and agriculture-related occupation. Only one study has reported about the adverse health effects among retail pesticide shopkeepers (Kesavachandran, Pathak, Fareed, *et al.* 2009a; Kesavachandran, Pathak, Fareed, *et al.* 2009b). The systematic review and meta-analysis and the pooled analysis of data from studies performed as per the evidence-based medicine techniques on adverse health incidences such as cancer, diabetes, neurodegenerative diseases, congenital abnormalities, and so on were unable to correlate with pesticide exposure. There is a need for more studies with better methodology, larger population size, and good epidemiological practices to gain more insights into the pesticide hazard in future.

In developed countries, poisoning from pesticide exposure among agricultural workers has been successfully reduced to a considerable extent by the use of PPE at workplace. The measures such as wearing PPE while handling pesticides and providing subsidized or free PPE through government scheme can help reduce the problem of pesticide toxicity considerably. In the developing countries, there should be strict guidelines for safe disposal of pesticide waste and containers. To avoid the indiscriminate use of pesticides, more awareness programmes about the safe use of pesticides may be conducted in rural areas. Regular monitoring of environmental matrix (air, water, soil, and food materials, such as fruits, cereals, and pulses) can help understand the toxicity levels and enable proper implementation of measures for protection against pesticide toxicity.

Periodic monitoring of clinical and subclinical activities, including biomarkers of pesticides, should be undertaken through subsidized or free health insurance schemes. The monitoring and assessment should be undertaken regularly at pesticide industries and retail pesticide shopkeepers for adequate health management. Facilities such as cholinesterase and pesticides estimation in blood should be made available in rural health care centres or hospitals which are vulnerable to pesticide exposure. More innovative tools for pesticide detection in human and environmental matrices should be developed considering the existing limited facilities to estimate pesticides and their metabolites using equipment, such as gas chromatography and high pressure liquid chromatography (HPLC) methods. Longitudinal cohort studies on agriculture workers should be undertaken in developing countries like India in national toxicology oriented research centres for better understanding of the problem.

Evidence-based agriculture or agricultural farm practices are not followed while spraying pesticides in agricultural farm, especially in developing countries like India. Agricultural workers handling pesticides should be given proper guidance and supervision by agriculture department for spraying of pesticides for specific crops, vegetables, and fruits. Retail markets of pesticides should be controlled by the agriculture department of a state government so that spurious pesticides may not enter the market. This will also reduce the large-scale stocking of pesticides by farmers, which often results in pesticides poisoning cases (e.g., suicides in farming community, accidents of poisoning among children of farmers).

REFERENCES

Agricultural Legislations. Details available at http://www.icar.org.in/files/Agril-Legislation.pdf, last accessed on 25 June 2011

Bhushan, P. 2011. The pesticides management bill can be pesticide (production and distribution) management bill, 2008. *Agri Activism,* 21 March. Details available at http://indiafarm.wordpress.com/2011/03/21/the-pesticides-management-bill-can-be-pesticide-production-and-distribution-management-bill-2008/, last accessed on 23 February 2012

Edwards, J. G. 2004. DDT: a case study in scientific fraud. *Journal of American Physicians and Surgeons* 9(3), 83–88

Factories Act in India. 2011. Details available at http://raghubar.wordpress.com/2010/10/16/189/, last accessed on 15 December 2011

Hagedorn, K. and R. Chennamaneni. 2011. Pesticide, residues and regulations in India: a case of vegetables in Hyderabad market. Research Report for Analysis and Actions for Sustainable Development of Hyderabad, Centre for Sustainable Agriculture. Details available at http://www.kisanswaraj.in/wp-content/uploads/pesticides-residues-regulation-CSA.pdf, last accessed on 21 December 2012

India Juris. 2011. Pesticides—Laws, Registration and FDI in India. Details available at http://www.indiajuris.com/pest.pdf, last accessed on 15 November 2011

Kesavachandran, C., M. K. Pathak, Md. Fareed, V. Bihari, N. Mathur, and A. K. Srivastava. 2009a. Health risks of employees working in pesticide retail shops: An exploratory study. *Indian Journal of Occupational and Environmental Medicine* 13(3): 121–126

Kesavachandran, C., M. Fareed, M. K. Pathak, V. Bihari, N. Mathur, and A. K. Srivastava. 2009b. Adverse health effects from pesticide exposure in agrarian populations of developing countries. *Reviews of Environmental Contamination and Toxicology* 200: 33–52

Ministry of Agriculture. 2011. Norms or guidelines for strengthening and modernisation of pest management approach in India. New Delhi: Department of Agriculture and Cooperation, Ministry of Agriculture, Government of India. Details available at http://agricoop.nic.in/smpma21910.pdf, last accessed on 1 July 2011

Misra, S. S. 2011. Pesticide management bill pushed for winter session. *Center for Science and Environment, October 2011.* Details available at http://www.cseindia.org/node/3189; last accessed on 14 January 2012

Mistry, K. U. 2011. Summary of Industrial Safety Laws in Gujarat, India. Details available at http://www.chemicalsafety.co.in/safety-law.pdf, last accessed on 1 September 2011

Mueller, T. 2011. Indian MRL: Overview and Trends. *California Speciality Crop Council, MRL Workshop*, San Francisco, 1–2 June 2011. Details available at http://specialtycrops.org/MRL_pdfs/PPT/7%20Indian%20MRLs%20Mueller.pdf

Pesticide Management Bill, 2008. Details available at http://www.cseindia.org/category/thesaurus/pesticide-management-bill-2008, last accessed on 3 March 2012

Planning Commission. 2011. Existing set-up of occupational safety and health in the workplace. Details available at http://dgfasli.nic.in/working_group/chap_1.htm, last accessed on 13 November 2011

Press Trust of India. 2010. Cabinet Nod to Bill for Safety of Plantation Workers. *The Hindu*, 15 March 2010. Details available at http://www.thehindu.com/news/national/article245202.ece?service=mobile, last accessed on 12 March 2012

Singh, M. 2011. Proceedings of the Asia Regional Workshop on the Implementation, Monitoring and Observance of the International Code of Conduct on the Distribution and Use of Pesticides. *FAO Corporate Document Repository.* Details available at http://www.fao.org/docrep/008/af340e/af340e0a.htm, last accessed on 5 December 2011

Shan, Z. J. 1997. Status of pesticide pollution and management of China. Environmental Protection 7: 40-43

Tata Chemicals Limited. 2009. Pesticide regulations in India. Indian Chemical Industry, D&B Sectoral Round Table Conferences Series. Details available at http://www.dnb.co.in/Chemical_2010/pestiChapter3.asp, last accessed on 3 December 2011

The Plantations Labour (Amendment) Act, 2010. Details available at http://www.prsindia.org/uploads/media/Acts/The%20Plantations%20Labour%20(Amendment)%20Act,%202010.pdf, last accessed on 12 January 2012

The Prevention of Food Adulteration Acts and Rules. 2011. Details available at http://dbtbiosafety.nic.in/act/PFA%20Acts%20and%20Rules.pdf, last accessed on 29 May 2011

CHAPTER 8

GAP AREAS

GOOD AGRICULTURAL PRACTICES

In India, food production has increased substantially in the last decade because of the increased use of agrochemicals. Agrochemicals include fertilizer and pesticides, such as insecticides, herbicides and fungicides. The country's food production per hectare has doubled since 1950, which corresponds to the increased use of pesticides (FAO 1982). An in-depth research of agriculture practices is required in order to understand how agricultural production is impacted by pesticide use.

FATE OF PESTICIDES IN ENVIRONMENT

Some of the environmental concerns are common for both the developing and developed nations. These include the following (FAO 1982):

- Pesticide movement
- Persistence and uptake of pesticides in plants
- Pest resistance to pesticides
- Repeated application of the same pesticide is reported to enhance the growth of the related specific decomposing microorganisms
- Exposure of users and consumers to pesticides, their residues, or their by-products

Emphasis should be given to the above-mentioned aspects in order to assess the impact of pesticide exposure on the environment.

When pesticides are sprayed, a considerable amount of its vaporized form mixes with air and gets absorbed in particulate matter. In rural areas, the pesticide leaches into the groundwater and reaches drinking water facility such as tube well. The meteorological factors such as wind, temperature, rainfall, and relative humidity act as pesticide carriers. These factors are responsible for the spread of pesticides from agricultural fields to various locations. The agricultural pesticides also get introduced into the water treatment plants from nearby river/reservoir. Water treatment plants are generally not equipped to remove pesticides from water and thus pesticides contaminate the drinking water supply. Therefore, it is important to devise a management plan for drinking water supply in urban and rural areas in order to prevent pesticide contamination of water supplies.

EVALUATING EVIDENCE ON ENVIRONMENTAL HEALTH RISK: NEW PARADIGMS

To assess the adverse health effects of pesticide exposure, integration of evidence from a variety of sources, such as experimental studies, both in animals and humans, *in vitro* studies, and epidemiological research is needed. For this purpose, understanding of the sources, nature and levels of exposure to which humans may be subjected, the nature of the health outcome or toxic effect of pesticide and the mechanisms by which this might occur, the relationship between dose and response, and the variability and susceptibility of potentially exposed populations is required. Interpretation of results need caution on – methods used for determination of environmental and health impacts and the modulators that are likely to influence these impacts should be carefully evaluated. The potential future challenges such as protecting sensitive data and ensuring their privacy by scientists, managing the increased availability of computerized data, developing and employing sophisticated statistical approaches, and employing epidemiological methods to investigate mechanistic pathways and gene–environment interactions should be analysed (Rushton and Elliot 2003).

Valid data on health impacts of environmental exposure are prerequisites for arriving at any conclusion regarding various potential environmental causes of diseases (Rushton and Elliot 2003). In future, exposure assessment methods are likely to improve with advancement in the exposure biomarkers and molecular epidemiology. This will result in increased specificity and, hence, improve the ability to detect true difference in disease risk (Hulka, Wilcosky, and Griffith 1990). However, these studies suffer from limitations. They take a lot of time and incur heavy expenses, particularly in geographical studies, where location is used as a proxy for exposure (Elliot, Wakefield, Best, *et al.* 2000). In days to come, large new databases of morbidity will be available with the growing use and computerization in health care (Rushton and Elliot 2003).

The presence of multiple toxicants in varying concentrations in different compartments of the environment and the need to find attributable risks to pesticide exposure necessitate the requirement for detailed gene – and environment-wide association studies. This will also help policy makers in developing strategy and regulating the pesticides that are especially harmful to human beings.

MOLECULAR EPIDEMIOLOGY

Progress in molecular epidemiology has enabled the combined use of scientific disciplines of epidemiology and molecular toxicology to examine the interactions between genetic and environmental factors in the cause of a disease. Several studies have shown that high-level exposure to substances does not affect all individuals equally. This finding supports the theory that lays stress on the fact that susceptibility and resistance of an individual to diseases may be influenced by genetic factors (Schulte and Perera 1993). In order to examine mechanistic pathways and determine gene−environment interactions, there is a need for future research to include measurement of susceptibility.

ROLE OF SYSTEMATIC REVIEW AND META-ANALYSIS

Systematic review and meta-analysis methods are well-developed in the field of clinical research and they involve collation of the literature on a specific area, assessment of the extent and quality of studies performed, and provision of a compilation of results, often including quantitative estimation of risk estimates from combined studies (Rushton and Elliot 2003). Apart from providing transparency and reproducibility, these methods also ensure ease of updating. These methods help in identifying gaps in the knowledge base and areas for future research. Statistically, quantitative meta-analyses are more effective than single studies and also provide a framework for examining the possible sources of heterogeneity between studies (Blettner, Sauerbrei, Schlehofer, *et al.* 1999). However, these techniques are surrounded with controversies, especially with regard to the opportunity for bias and other sources of heterogeneity, in comparison to clinical trials. However, there is an increased use of these techniques in epidemiological research. Recently, numerous guidelines have been devised for their application (Egger, Schneider, and Smith 1998; Sutton, Abrams, Jones, *et al.* 2000).

In the past decade, the interest has shifted towards cross-design synthesis, that is, the quantitative combination of results from different study designs (Piegorsch and Cox 1996). The potential of meta-analysis techniques to merge data of both animal studies and human studies is being explored, and this will play a significant role in the development of risk assessments and setting of environmental standards (Rushton and Elliot 2003). Altered metabolic parameters among pesticide exposed population is another area of interest for future.

SOLUTIONS AND ROADMAPS FOR FURTHER RESEARCH

The following are the highlights of panel discussion on research needs for effective and safe use of pesticides in developing countries (FAO 1982):

• The research workers in developing countries are more engaged in carrying out fundamental research on pesticide's mode of action, metabolism, interaction, and environmental fate. In contrast, they should undertake adaptive research aimed at finding ways to utilize the existing pesticides effectively, efficiently, and safely.

• The needs and conditions of advanced countries are central to most existing data and information on pesticides, whereas prevailing needs and conditions of the developing countries should be considered while understanding future research.

• In advanced countries, radio-labelled metabolism studies are performed on almost every pesticide used in developing countries. It is imperative that those studies should be conducted in developing countries to examine the fate of pesticides under local conditions and in local crops.

• The farmers in developing countries should be trained to use pesticides safely and in accordance with the restrictions.

• Climatic conditions of tropics are different from the temperate zones. This difference is reflected in the pesticide's behaviour in the two different climatic zones. This calls for a different approach to examine the fate of the pesticides in both food and environment under both conditions.

HEALTH MANAGEMENT

The salient features of the gap areas in health management based on the studies performed by Ajayi, Akinnifesi, and Sileshi (2011) and van den Berg, Hii, Soares, *et al.* (2011) are as follows:

• It is important to assess and understand the relationship between pesticide usage and their resultant adverse effects on the human health, particularly in developing countries where regulations are poorly implemented and farmers' knowledge of safe handling and use of pesticides is inadequate.

- Although precautionary measures have been adopted to minimize exposure to pesticides, occupational pesticide poisoning still occurs in rural households and it has become a major concern in the development planning in agriculture.

- There is a need to increase the level of awareness and knowledge of households in order to minimize the hazards associated with the pesticide use in agricultural households.

- Encouraging pest management could be a complementary approach which can reduce the chemical use and exposure to occupational hazards among farm households. An example of this approach is the integrated pest management (IPM).

- Only a few cases of health symptoms related to pesticide poisoning are reported in health centres. This reflects the gaps in pesticide documentation. A mechanism for proper documentation of pesticide poisoning cases should be devised in order to develop better health management practices for vulnerable population.

- Better economic measures like providing incentives such as free medical assistance to all victims of pesticide poisoning cases in health clinics can be implemented. Reliable monitoring, assessment and reporting procedures regarding pesticide related health problems are necessary to formulate appropriate policies and regulations to minimize adverse health effects of pesticides.

- There are numerous gaps in pesticide management for vector control measures. Major gaps were generally observed in countries in pesticide procurement practices, training on vector control decision-making, certification and quality control of pesticide application, monitoring of worker safety, public awareness programme, and safe disposal of pesticide-related waste.

- Many countries do not pay due attention to effective and safe use of pesticides for vector control. This causes wastage of resources, suboptimal effectiveness of interventions, and harmful effects on environment and health of humans.

- The urgency of this situation is evident from the increased application of pesticides for control of malaria in recent times in countries that have intensified interventions in this regard. However, this development has not been followed by investment in pesticide management.

- Therefore, vector control programme should include capacity building on pesticide management and decision-making within the context of an integrated vector management (IVM) approach based on evidence. This should become a condition for support on vector control given by donors and funding agencies.

- Documentation of policies of public health pesticide management and agricultural pesticide management is necessary and amendments in existing legislations should be made or new legislations can be formulated and incorporated for strict implementation of management practices or guidelines for public health and agricultural sector.

REFERENCES

Ajayi, O. C., F. K. Akinnifesi, and G. Sileshi. 2011. Human health and occupational exposure to pesticides among smallholder farmers in cotton zones of Côte d'Ivoire. Details available at http://www.scirp.org/journal/HEALTH/ Vol.3, No.10, 631-637 (2011) doi:10.4236/health.2011.310107

Blettner, M., W. Sauerbrei, B. Schlehofer, T. Scheuchenpflug, and C. Friedenreich. 1999. Traditional reviews, meta-analyses and pooled analyses in epidemiology. *Int J Epidemiol* 28: 1 -9

Egger, M., M. Schneider, and G. D. Smith. 1998. Meta-analysis: spurious precision? Meta-analysis of observational studies. *BMJ* 316: 140 -4

FAO/IAEA. 1982. International symposium on agrochemicals: fate in food and the environment using isotope techniques, held in FAO Headquarters in Rome, Italy, 7 to 11 June 1982, *IAEA BULLETIN,* Vol 24, No.3

Hulka, B. S., T. C. Wilcosky, and J. D. Griffith. 1990. *Biological Markers in Epidemiology.* Oxford: Oxford University Press

Piegorsch, W. W. and L. H. Cox. 1996. Combining environmental information II: Environmental epidemiology and toxicology. *Environmetrics* 7: 309 -24

Rushton, L. and P. Elliot. 2003. Evaluating evidence on environmental health risks. *Br Med Bull* 68: 113 -128.

Schulte, P. A. and F. P. Perera. 1993. *Molecular Epidemiology Principles* and Practices. London: Academic Press, Inc.

Sutton, A. J., K. R. Abrams, D. R. Jones, T. A. Sheldon, and F. Song. 2000. *Methods for Meta-analysis in Medical Research.* Chichester: John Wiley & Sons Ltd

van den Berg, H., J. Hii, A. Soares, A. Mnzava, B. Ameneshewa, A. P. Dash, M. Ejov, S. H. Tan, G. Matthews, R. S. Yadav, and M. Zaim. 2011. Status of pesticide management in the practice of vector control: a global survey in countries at risk of malaria or other major vector-borne diseases. *Malaria Journal* 10: 125

INDEX

About the Authors

Anup Kumar Srivastava, MBBS, MD, FRIPHH, has successfully applied epidemiological tools for finding solutions to toxicological health problems, associated with the use of pesticides, metals, and cyclic compounds, among diverse populations. He has been actively involved in research for the past 35 years, and has worked at CSIR-Indian Institute of Toxicological Research (CSIR-IITR), Lucknow. Dr Srivastava also served as Director of National Institute Miners' Health, Nagpur and was the first specialist of Industrial Health appointed in the coal industry at Western Coalfields Limited, Nagpur.

Mashelkar Committee on Auto Fuel Policy entrusted him with the responsibility of appraising the health impact of air pollution with reference to vehicular emissions. He has successfully assisted many national and international committees on environmental health issues. He has extensively researched and published in the discipline of environment and occupational health. He retired as Head, Epidemiology Section, CSIR-IITR in 2013 and is currently working with Hind Institute of Medical Sciences, Lucknow.

C Kesavachandran, PhD, is Senior Scientist, Epidemiological Section, CSIR-IITR. He received his PhD from Mahatma Gandhi University, Kerala and is pursuing a research career as a Scientist at the CSIR-Indian Institute of Toxicology Research (CSIR-IITR), Lucknow, since 2002. Previously, Dr Kesavachandran worked as a Guest Scientist at Centre for Excellence in Epidemiology (CV care) at University Clinics Eppendorf, Hamburg, Germany from 2011 to 2012. He has been actively engaged in research in many key areas, such as occupational and environmental health, exercise physiology, nutritional epidemiology, physiological mechanisms of toxicants in respiratory system, meta-analysis, industrial hygiene studies, environmental monitoring of pollutants and its health effects. He has more than 50 publications in peer-reviewed national and international journals.